International Trade in Forest Products: Lumber Trade Disputes, Models and Examples

International Trade in Forest Products: Lumber Trade Disputes, Models and Examples

EDITED BY

G. Cornelis van Kooten and Linda Voss

Department of Economics
University of Victoria
Victoria, Canada

CABI

CABI is a trading name of CAB International

CABI
Nosworthy Way
Wallingford
Oxfordshire OX10 8DE
UK

Tel: +44 (0)1491 832111
Fax: +44 (0)1491 833508
E-mail: info@cabi.org
Website: www.cabi.org

CABI
WeWork
One Lincoln St
24th Floor
Boston, MA 02111
USA

T: +1 (617)682-9015
E-mail: cabi-nao@cabi.org

A catalogue record for this book is available from the British Library, London, UK.

Library of Congress Cataloging-in-Publication Data

Names: Van Kooten, G. C. (Gerrit Cornelis), editor.
Title: International trade in forest products : lumber trade disputes, models and examples / edited by G. Cornelis van Kooten and Linda Voss, Department of Economics, University of Victoria, Victoria, Canada.
Description: Wallingford, Oxfordshire ; Boston, MA : CABI, [2021] | Includes bibliographical references and index. | Summary: "This book provides the background theory behind two types of trade models rooted in economic theory and mathematical programming, and two econometric/statistical models. It shows the reader how to construct their own models to analyse policy questions in forestry and other sectors"-- Provided by publisher.
Identifiers: LCCN 2020036407 (print) | LCCN 2020036408 (ebook) | ISBN 9781789248234 (hardback) | ISBN 9781789248241 (ebook) | ISBN 9781789248258 (epub)
Subjects: LCSH: Forest products industry--United States. | Forest products industry--Canada. | United States--Commerce--Canada. | Canada--Commerce--United States. | International trade--Econometric models.
Classification: LCC HD9764.U62 I58 2021 (print) | LCC HD9764.U62 (ebook) | DDC 382/.456741440973--dc23
LC record available at https://lccn.loc.gov/2020036407
LC ebook record available at https://lccn.loc.gov/2020036408
References to Internet websites (URLs) were accurate at the time of writing.

ISBN-13: 9781789248234 (hardback)
9781789248241 (ePDF)
9781789248258 (ePub)

Commissioning Editor: Ward Cooper
Editorial Assistant: Lauren Davies
Production Editor: James Bishop

Typeset by SPI, Pondicherry, India
Printed and bound in the UK by CPI Group (UK) Ltd, Croydon, CR0 4YY

Contents

List of Contributors

Joseph Buongiorno is the Class of 1933 Bascom Professor Emeritus & John N. McGovern WARF Professor Emeritus in the Department of Forest and Wildlife Ecology, University of Wisconsin-Madison. Email: jbuongio@wisc.edu

Jinggang Guo is a Researcher at RTI International, Research Triangle Park, NC. Email: jinggang.guo@outlook.com

Craig Johnston is a senior economist with the Commodities Division in the International Economic Analysis Department with the Bank of Canada. Email: CJohnston@bank-banque-canada.ca

Xintong Li is a graduate student of the Department of Economics at the University of Victoria, Canada. Email: xintongl@uvi.ca

Fatemeh Mokhtarzadeh is a Post-doctoral Researcher in the Department of Economics at the University of Victoria, Canada. Email: fatemehm@uvic.ca

Harry Nelson is an Associate Professor in the Department of Forest Resources Management at the University of British Columbia, Canada. Email: harry.nelson@ubc.ca

Prakash Nepal is a Research Economist at the USDA Forest Products Laboratory. Email: prakash.nepal@usda.gov

Jeffrey Prestemon is a project leader at the U.S. Forest Service, Southern Research Station and an Adjunct Professor in the Department of Forestry and Environmental Resources at North Carolina State University. Email: jeff.prestemon@usda.gov

Brad Stennes is a director and senior economist at Canadian Forest Service's Pacific Forestry Centre in Victoria, BC. Email: bstennes@pfc.forestry.ca

G. Cornelis van Kooten is Professor and Canada Research Chair in the Department of Economics at the University of Victoria, Canada. Email: kooten@uvic.ca

Linda Voss is a research assistant in the Department of Economics at the University of Victoria, Canada. Email: repa@uvic.ca

Acknowledgements

The editors acknowledge funding support from Canada's Social Sciences and Humanities Research Council's Partnership Development Grant program, SSHRC #890-2016-0064.

Material in Chapter 3 is based on Johnston, C. (2014) Topics in forest product modelling: the economics of bioenergy product exports from forests. Unpublished PhD dissertation. Department of Economics, University of Victoria, Victoria, Canada.

The authors of Chapter 6 are grateful for financial support over the years from the Food and Agriculture Organization of the United Nations, the USDA Forest Service Forest Products Laboratory, the USDA Forest Service Southern Research Station, and the University of Wisconsin-Madison; they are also extremely grateful to Dr Csaba Mészáros for allowing them to use the interior point quadratic optimizer BPMPD as the GFPM solver.

The research leading to Chapters 3, 4 and 6, was supported in part by joint venture agreements between the University of Wisconsin (16-JV-11330143-039 and 17-JV-11330143-087) and the United States Department of Agriculture Forest Service Southern Research Station (20-IJ-11330180-050)

The views presented in chapters 3, 4 and 6 are those of the authors and do not represent the Bank of Canada's position.

All errors are the authors' own.

SSHRC CRSH

1 Introduction

G. Cornelis van Kooten

University of Victoria, Victoria, BC Canada

Abstract

In this chapter, we provide a brief background to the major players in global lumber markets, both countries and companies. In addition, the four measures economists use for measuring welfare and income redistributions are described, and the remaining chapters outlined.

The Canada–United States softwood lumber dispute has a history that goes back to the late 19th century. As with many disputes related to forestry and forest product, this dispute relates primarily to institutional differences in the way Canada and the U.S. govern forestlands. In Canada, the vast majority of forestland is publicly owned, with provinces having been granted ownership over forest resources by the Constitution. As a result, provincial governments became owners of all forestlands within their jurisdiction, with some exceptions related to land under the purview of the federal government (viz. National Parks and aboriginal lands) and lands granted to the Canadian Pacific Railway in exchange for completing construction of a transcontinental railway. This constitutional arrangement is discussed in greater detail in Chapter 2.

Although there is nothing that prevented provinces from selling forestland to private holders, the public has generally opposed such sales. With the exception of lands that fell into private hands through outright sale or as a grant, transfers of public forestland to the private sector ended in the early 1900s (Marchak, 1993, 1995, pp. 85–116; Cashore, 1997). Such transfers would now be opposed by the majority of citizens who attach value to the non-commercial attributes of the land – to wildlife habitat, recreational and visual amenities, and environmental services (e.g. carbon sink, absorber of pollutants, water regulation, etc.), to name a few.

Federal and state governments in the U.S. also own vast tracts of forestland, mainly in the West, but most of the better, more productive lands are in private

DOI: 10.1079/9781789248234.0001

hands. Because more logs are harvested from private than public lands in the U.S., log prices are relatively transparent. A sawmill would have some notion of what price it would be willing to pay for logs based on market-determined prices of lumber and other wood products at the time. Logging companies compete for the right to harvest logs on both private and public lands, with stumpage prices determined through a bidding process that takes into account current market conditions for wood products.

In Canada, on the other hand, stumpage prices of logs are set by government and are not sensitive to lumber and other wood product prices. Under a variety of tenure arrangements, discussed in Chapter 2, Canadian sawmills access logs at whatever fee is set by the province. If the price is too high, fewer logs might be harvested; if the price is low, more logs might be harvested. The word 'might' is used to reflect regulations on the volume of timber that a tenure holder is required to harvest in a given year or over some government-approved planning period (usually 5 years). This friction between what is thought of as an open market for logs (U.S.) and a regulated market for logs (Canada) is a main reason for the softwood lumber dispute. Forestlands and forest practices in both Canada and the U.S. are affected by public attitudes to forest conservation, the need to certify sustainable forest practices (or lack thereof), tenure arrangements, governance, and other social controls. Similar factors affect forest practices and production of forest products across the globe.

In this book, the main focus is on one aspect of the forest sector, namely, how the softwood lumber dispute affects bilateral trade flows between Canada and the U.S., and between North America and the rest of the world. In doing so, we also gain some knowledge about the impacts of trade distortions on economic wellbeing in various regions, and which groups in society might gain or lose, and by how much. However, the models we consider in the chapters that follow can be used to investigate various policies related to forest sector trade; indeed, we provide examples of how the models can be expanded to examine log export restrictions, subsidies on wood biomass for generating electricity, impacts of climate change, and so on. The Canada–U.S. lumber dispute is only one example of the types of policies that can be investigated using trade models, albeit an important example.

As further background to the softwood lumber trade dispute, consider the main countries that produce and export lumber. This is provided in Table 1.1. Not surprisingly, the main players are the U.S., Canada, Russia, the Scandinavians, China, and several central European countries. The top four producers are the U.S., Canada, Russia, and China, and they accounted for 47% of global production of lumber in 2018; the top four exporters are Russia, Canada, Scandinavia, and Germany/Austria, which together accounted for 73% of exports in 2018. Canada and Russia each account for about 23% of the world's total exports of softwood lumber, with the Scandinavians contributing about 16%.

This concentration of production and exports is reflected at the company and sawmill level. 'The Sawmill Database' currently indexes some 1235 sawmills and 700 companies in 43 countries; these sawmills produce some 164 million m^3 of softwood lumber each year, which represents approximately 44% of global production. The top 12 lumber-producing companies and their annual

Table 1.1. Top 12 countries producing and exporting lumber, 2018 (m^3 per year).

Production (m^3)			Exports (m^3)		
Rank	Country		Rank	Country	
1	United States	59,458,000	1	Russian Federation	29,754,000
2	Canada	45,227,520	2	Canada	29,655,922
3	Russian Federation	39,492,000	3	Sweden	12,437,000
4	China	38,361,000	4	Finland	8,685,340
5	Germany	22,609,724	5	Germany	8,223,141
6	Sweden	18,273,000	6	Austria	5,918,869
7	Finland	11,800,000	7	Belarus	3,345,800
8	Austria	10,221,000	8	Ukraine	3,188,800
9	Japan	8,606,000	9	Chile	3,099,737
10	Brazil	8,600,000	10	United States	2,880,000
11	Chile	8,025,300	11	Latvia	2,764,000
12	France	6,553,000	12	Brazil	2,374,000
	Global	**390,226,847**		**Global**	**129,866,876**

Adapted from: FAO database at http://www.fao.org/faostat/en/#data/FO (accessed June 22, 2020)

Table 1.2. Top twelve global producers of lumber by production or capacity (m^3 per year).

Rank	Company[a]	Capacity[b]	Rank	Sawmill[c]	Capacity[b]
1	West Fraser Timber Co Ltd (Canada)	8,460,000	1	Klausner Holz Thüringen	1,200,000
2	Canfor (Canada)	6,900,000	2	Wismar Sawmill	1,200,000
3	Weyerhaeuser (U.S.)	6,449,000	3	Plateau sawmill	1,054,000
4	Stora Enso (Finland)	4,646,000	4	Houston sawmill	1,020,000
5	Georgia Pacific (U.S.)	4,300,000	5	Quensel West Fraser	940,000
6	Resolute Forest Products (Canada)	4,000,000	6	Binderholz sägewerk	825,000
7	Interfor (Canada)	3,550,000	7	Mackenzie sawmill	800,000
8	Sierra Pacific Industries (U.S.)	3,200,000	8	Landsberg Am Lech[b]	800,000
9	Hampton Affiliates (U.S.)	3,100,000	9	Offner Wolfberg[c]	800,000
10	Arauco (Chile)	2,800,000	10	Dunkley sawmill	800,000
11	Tolko Industries Ltd (Canada)	2,500,000	11	Baur Holz[b]	750,000
12	Schweighofer (Austria)	2,400,000	12	Longview Softwood[d]	720,000

[a] Country of head office indicated in parentheses.
[b] Capacity is employed, but in some cases this refers to actual production.
[c] Mills 1, 2, 8 and 11 are in Germany; 6 and 9 in Austria; 12 in the U.S.; and 3, 4, 5, 7 and 10 in Canada
Adapted from: The Sawmill Database https://www.sawmilldatabase.com/productiontoplist.php (accessed June 16, 2020).

production capacity are found in Table 1.2, as are the top 12 producing sawmills. Five of the top 12 forestry companies are Canadian, four are American, one is Finnish, one is Chilean, and one is Austrian. Surprisingly, no German forest company is ranked in the top 12 global lumber producers, although four of the largest sawmills are located in Germany, and Germany is the fifth largest producer and exporter of lumber.

Our interest is in the economics of wood products production and trade. Economists have the tools required to analyze public policies – to determine whether or not a policy enhances society's overall wellbeing (or welfare), and identifying which groups in society gain and which lose from the policy, and the potential size of such gains or losses. Although the welfare measures are well known from a theoretical standpoint, economists apply a variety of empirical methods to calculate the various welfare measures. These are briefly reviewed below.

1.1 Welfare Measures

In economics, costs and benefits constitute a surplus that is either lost (cost) or gained (benefit). There are four types of economic surplus.

1. Consumer surplus is the difference between what consumers are willing to pay for goods and services and the actual expenditure required to obtain those goods and services. In essence, it is the difference between the total benefit that consumers derive from purchasing an amount Q and what they actually have to pay (which equals price times quantity, $P \times Q$). It can be measured by the area below the marginal willingness-to-pay (demand) function minus what they actually pay, and can be represented mathematically as:

$$\text{CS} = \int_0^Q \left[D(q) - S(q)\right] dq, \quad (1.1)$$

where CS refers to the consumer surplus. The integral constitutes the total willingness to pay (area under demand function) minus the total cost of production (area under supply function, or $P \times Q$). The consumer surplus is illustrated in Fig. 1.1.

Consumer surplus is not always directly measurable. Consider the case where a project does not affect consumer surplus because the market price is unaffected. For example, it is unlikely that decisions concerning the harvest or protection of a single commercial forest landscape will affect the prices of timber products. In that case, the direct consumer surplus associated with such a project is unlikely to change; indeed, unless the project lowers (raises) price, the consumer is not going to gain (lose) surplus from the project. In that case, consumer surplus becomes relevant only in some other market, but not the market for lumber. Suppose that, in addition to the market for lumber, there is a demand for an environmental amenity that is somehow impacted by a logging decision. In that case, there may be surplus that needs to be taken into account in evaluating the logging project. This would be an indirect cost or benefit associated with the project, which is discussed below as the fourth type of surplus.

2. Quasi-rent constitutes the difference between total revenue and total variable cost. It is measured by the area below price and above the marginal cost (supply) function, as indicated in Fig. 1.1. It is also referred to as producer

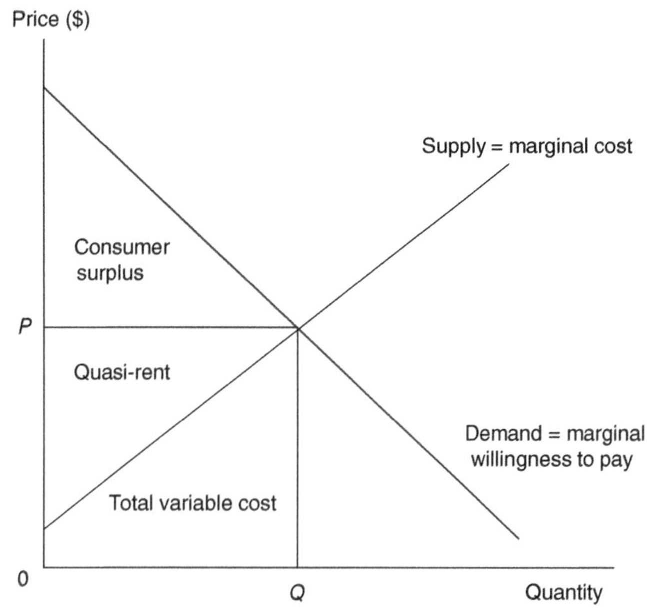

Fig. 1.1. Consumer and producer surplus. (Author's own figure.)

surplus. While constituting a true welfare measure, quasi-rent (producer surplus) accrues to fixed factors of production and entrepreneurship. That is, in Fig. 1.1, the supply curve is a short-run supply function, which means that returns to the fixed factors of production must come from quasi-rent. Any attempt to tax this quasi-rent or producer surplus will disincentivize firms from investing in the production of that commodity.

Consumer surplus and quasi-rent can be considered from a perspective that ignores the utilitarian foundations of demand. Think of the supply function as the minimum amount that a seller is willing to accept (WTA), at the margin, to part with the item in question. Any amount above minimum WTA constitutes a surplus to the seller. Likewise, the demand function can be thought of as the maximum amount a buyer is willing to pay (WTP). Suppose that person A is willing to sell an article at any price above $9.50, while person B is willing to pay upwards of $17.50 for that item. If they agree to a price of $12, person A gains a surplus of $2.50 while B gains a surplus of $5.50, so the total surplus shared between them is $8. If the trade had occurred at $15, A would have realized a surplus of $5.50 and B a surplus of $2.50, but again the benefit to society (the two people) is still $8. Even if A was coerced to sell at $8, so that B gained $9.50, the total surplus of this exchange would remain $8 (= $9.50 – $1.50).

3. Resource rent accrues to natural resources and consists of two components that are often indistinguishable from each other in practice, and difficult to separate from the second type of surplus – the quasi-rent (producer surplus). We illustrate the concept of resource rent with the aid of Fig. 1.2. The supply curve in this figure is a long-run supply curve, and differs from the short-run supply curve in Fig. 1.1 because, in the long run, all factors of production are

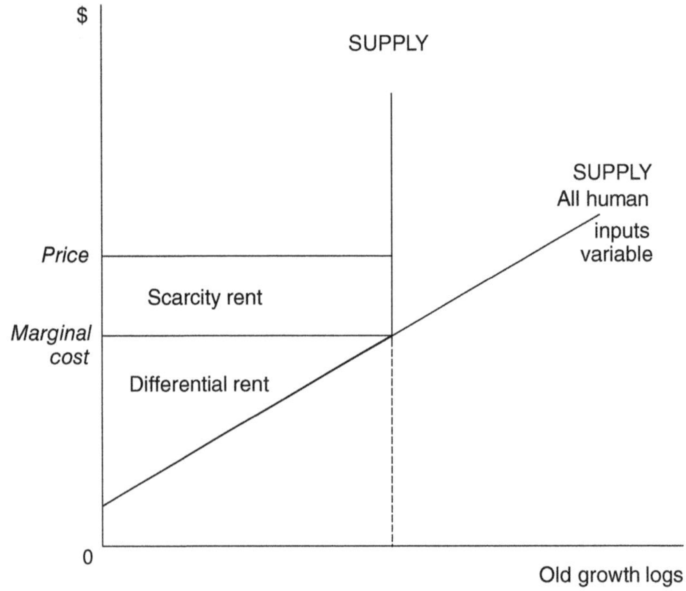

Fig. 1.2. Scarcity and differential components of resource rent. (Author's own figure.)

variable. In that case, there is no quasi-rent to be measured, because quasi-rent accrues only to fixed factors of production, of which there are none in the long run, except natural resources.

The first component of resource rent is **differential** (sometimes referred to as **Ricardian**) **rent** that arises because of inherent or natural advantages of one location relative to another. For example, a woodlot located near a sawmill or transportation corridor (highway, water transport) earns a windfall relative to an identical woodlot that is located farther from a sawmill or transportation corridor.

A second component is referred to as **scarcity rent**, which results simply because there is a limit to the number of stands with commercial timber. That is, if the timber producer, despite being the highest cost producer in the world, earns a windfall over and above what could be earned elsewhere in the economy, there is a scarcity rent because price exceeds the marginal cost of production. Given that old-growth timber that produces large logs for decorative use is limited, the supply curve is vertical at that limit, thereby creating a scarcity rent.

Resource rent is the sum of the differential and scarcity rents, and is often the greatest benefit in decisions about whether to harvest a forest. The government can tax resource rents without adversely affecting private investment decisions. However, because the measurement of resource rents is difficult, governments must be careful in taxing such rents lest quasi-rents be taxed instead. If quasi-rents are taxed, returns to investors are insufficient to cover fixed costs and fewer log trucks and processing plants are built.

4. Finally, the **indirect surplus** refers to benefits or costs that accrue in markets for substitute and/or complementary goods and services. However, indirect benefits occur only if price exceeds marginal cost in one of the affected markets (Boadway and Bruce, 1984, pp. 252–255). Whenever price exceeds marginal cost, this implies that society values the good or amenity more than it costs to provide it. Hence, if there is an increase in the demand for the good so that more of the good or amenity is purchased, there is a benefit given by the difference between price (marginal willingness to pay) and marginal cost (willingness to accept) for each unit purchased; this occurs until the declining marginal willingness to pay equates to a rising marginal cost as more is produced. Conversely, there is a cost if the opposite is true – demand shifts inward and less is bought. Note that there is an indirect benefit only if price exceeds marginal cost and individuals are free to increase or decrease the quantity they purchase; if not, there is nothing to measure in the market for substitutes or complements. Note that, if price equals marginal cost in each of the markets for substitutes and complements, there are no indirect effects even if the amount purchased changes.

1.2 Outline of Remainder of Book

The purpose of this study is to explain the various models that economists have used to study the impacts of trade distortions in global forest products trade on commodity availability, price, and wellbeing in various regions. The main driver behind much of the modeling effort in the forestry sector has been the Canada–U.S. softwood lumber trade dispute. Yet, despite this focus on Canada–U.S. lumber trade, models have expanded to include many other forest products and jurisdictions outside of North America.

Nonetheless, we begin in Chapter 2 by examining the history of the Canada–U.S. softwood lumber dispute, which has its origins in the different tenure arrangements between the countries – that the degree of public ownership of forestlands in Canada is much greater than that in the U.S. Canadian tenure arrangements are considered to lead to stumpage prices that are not market driven and, therefore, are a potential mechanism for subsidizing downstream lumber and other wood product producers. Price setting is opaque because prices are determined by fiat. In the U.S. timber from public forestlands is sold by auction; but, since more timber originates with private land, U.S. prices are essentially set by market forces. Interestingly, while the U.S. Forest Service is part of the U.S. Department of Agriculture (USDA), agricultural commodity prices are set by Congress – by fiat (Schmitz *et al.*, 2020).

Different approaches for modeling trade in forest products are also discussed in Chapter 2. Spatial price equilibrium (SPE) trade models are perhaps the main tool used to examine the impacts of various trade restrictions on the production, consumption, and bilateral trade flows of various forest products, as well as effects on prices and wellbeing in various jurisdictions, where wellbeing takes into account the consumer and producer surpluses plus any tax revenues and/or policy-induced rents. Following on the welfare measures described earlier in this chapter and the brief discussion of modeling approaches in the latter half of Chapter 2, the economic theory underlying the measurement of wellbeing across vertically and horizontally chained product markets, and the development of SPE trade models, are described in Chapter 3.

Two general approaches have been used to construct SPE models. One approach is to examine direct bilateral trade flows between regions. This approach is used in what we will refer to as the Resource Economics and Policy Analysis (REPA) forest products trade model. The REPA trade model originates with the University of British Columbia, the University of Victoria, and the Pacific Forest Centre of the Canadian Forest Service in Victoria. The REPA SPE model is not fixed, in the sense that the model changes as necessary to answer particular questions. The structure of the model is such that the number of regions in the model is fluid, while the method used to solve the model also varies from one application to another. For example, the objective function in the REPA model is quadratic, so that it can be solved using a quadratic programming algorithm that usually relies on the simplex method. However, if an ad valorem tariff is levied by some jurisdiction, then price and quantity need to be determined simultaneously, so models are generally solved using a linear mixed complementarity programming (MCP) algorithm (see Chapter 5, this volume).

In Chapter 4, the basic structure of the multi-region, multi-product trade model is described, along with a discussion of how the model can be calibrated using positive mathematical programming (Howitt, 1995). Calibration allows us to replicate existing bilateral trade flows. Once the model is calibrated, it can then be used to analyze policy questions. Several versions of the REPA trade model are also discussed, while the General Algebraic Modeling System (GAMS) code and data for running these basic models are provided in Appendices to the chapter. These models include a version that employs a two-stage calibration procedure.

If sufficient data are available, the REPA model can be modified to address many policy questions that relate to the forest sector and trade. Therefore, to illustrate the use of the REPA trade model and how it can be modified, it is used to investigate several policies related to the Canada–U.S. softwood lumber dispute in Chapter 5. One such policy is the use of a quota rather than an export tax or import tariff to restrict trade. In addition to a discussion of these policy options, further GAMS code and data are provided in the Appendices to Chapter 5.

An alternative approach to constructing an SPE model is found in Chapter 6, which discusses the Global Forest Products trade Model (GFPM) that originated with Joseph Buongoriono at the University of Wisconsin at Madison. Rather than examining bilateral trade flows between the various countries in the model, GFPM has countries trading with a central exchange, resolving bilateral trade flows after the fact. Note that the model is country-based rather than region-based, although trade flows to regions within a country can also be resolved ex post. Because of its structure, GFPM does not have the same 'looseness' of structure as the REPA model – it is an off-the-shelf model that other users can use without too much reprogramming. It also handles a much richer set of wood product categories. A description of GFPM is provided in Chapter 6, while the GFPM 2017 software and latest data are available online at https://buongiorno.russell.wisc.edu/gfpm/ (accessed September 4, 2020).

In addition to SPE models, economists have used various types of econometric models to study trade flows, specifically gravity models and a Global Vector Autoregressive (GVAR) approach. A gravity model employs the idea from physics that (astronomical) bodies are attracted to each other on the basis of their mass. Likewise, economies with a higher per capita gross domestic product (GDP) attract greater imports than those with less per capita GDP. Since tariff and non-tariff barriers are an obstacle to trade, a statistical trade model can incorporate these policies as explanatory variables in a regression equation along with the gravity regressors. As an illustration of this approach, the development of a gravity model for Canada–U.S. trade is described and estimated in Chapter 7.

Finally, the GVAR approach employs time series analysis to analyze trade flows, and is described in Chapter 8. It has the advantage of relating trade in forest products to external factors such as oil prices or GDP. This enables the analyst to examine the impact that a shock to oil prices will have on future trade. The GVAR approach can also be used to investigate the effect that a shock, such as the COVID-19 pandemic, can have on forest product prices. Along with the gravity modeling approach, GVAR constitutes an econometric approach to the analysis of forest products trade that is rooted in observed behavior and is meant to complement the SPE models of Chapters 4 through 6.

One alternative not discussed in this study is the use of computable general equilibrium (CGE) approaches. One reason is that the forest sector, while important to some economies, is a very small component of the global economy, contributing less than 1% to global GDP. Further, and related to this point, the data required to construct a global CGE model are not sufficiently disaggregated to permit examination of various forest products; rather than facilitating investigation of softwood lumber trade, for example, a CGE model would enable one to consider the entire forestry sector as one component of the model, or, more likely, forestry combined with agriculture. CGE models do not permit the detail required to examine individual forest products trade, unlike the models considered in the current work. Some discussion of this is found in the concluding Chapter 9.

References

Boadway, R.W. and Bruce, N. (1984) *Welfare Economics*. Blackwell Publishing, Oxford, UK.

Cashore, B. (1997) Flights of the phoenix: explaining the durability of the Canada–U.S. softwood lumber dispute. *Canadian–American Public Policy* 32, December, 1.

Howitt, R.E. (1995) Positive mathematical programming, *American Journal of Agricultural Economics* 77, 329–342.

Marchak, P.M. (1993) *Green Gold: The Forest Industry in British Columbia*. University of British Columbia Press, Vancouver, Canada.

Marchak, P.M. (1995) *Logging the Globe*. McGill-Queen's University Press, Montreal and Kingston, Canada.

Schmitz, A., Moss, C., Schmitz, T.G., van Kooten, G.C. and Schmitz, H.C. (2020) *Agricultural Policy, Agribusiness and Rent Seeking Behavior*, 3rd edn. University of Toronto Press, Toronto, Canada (in press).

2 Canada–U.S. Softwood Lumber Dispute: Background

G. Cornelis van Kooten[1], Harry Nelson[2] and Fatemeh Mokhtarzadeh[1]

[1]*University of Victoria, Victoria, BC Canada;* [2]*University of British Columbia, Vancouver, BC Canada*

Abstract

In this chapter, we examine the importance of softwood lumber production to Canada's economy and provide a brief history of the Canada–U.S. softwood lumber dispute and its resolution on various occasions using U.S. countervailing and anti-dumping duties, export taxes or various types of quota regimes, including tariff rate quotas. The construction of excess supply and demand functions is explained, as are the gains from trade. This helps inform the modeling approaches that are identified in later chapters.

Canada is one of the largest producers and exporters of softwood lumber in the world. Its forest industry employs more than 200,000 people (including 12,000 indigenous people), accounts for 7.2% of total exports, generates nearly $1.5 billion in government revenues, and contributes about $25 billion to the economy (1.6% of Canadian GDP) (Natural Resources Canada, hereafter NRC, 2018). In 2017, softwood lumber accounted for 29% of total forest product exports, followed by wood pulp (23%), newsprint and printing paper (12%), and structured wood panels used in construction (7%), with the remainder constituting a variety of products. Most of the non-lumber forest products that are produced depend on residues from sawmilling. Historically, the United States has been Canada's major export market for softwood lumber (SWL), and, given the size of its domestic market, the U.S. is likely to continue to be heavily reliant on SWL imports from Canada (NRC, 2017).

Softwood lumber derives from cedar, Douglas fir, pine, and spruce. It is broadly used in North America and elsewhere in construction and the making of furniture and other wood products. The softwood lumber industry in both countries depends on the U.S. housing market, particularly new construction and renovations. Prior to the 2007/2008 financial crisis, for example, softwood lumber sales to the U.S. prospered as the U.S. residential real-estate market boomed. The proportion of Canadian lumber exports destined for the U.S.

10.1079/9781789248234.0002

peaked at about 80% in 2005/2006, with export volume reaching a high of about 50 million cubic meters (m^3) and subsequently falling to its lowest level in 2011 as a result of the U.S. housing market crisis (see Figs 2.1 and 2.2). The U.S. share of Canadian lumber exports rose again thereafter as the U.S. housing market began a modest recovery. Nonetheless, some 60% or more of Canadian lumber exports depend on the U.S. market (Fig. 2.2), despite growing exports of softwood lumber to China over the past decade – to about 14% of total exports in 2016. Also identified in Fig. 2.2 are periods of note in the Canada–U.S. softwood lumber dispute (discussed in more detail in section 2.2).

The monthly average of the Random Lengths' Framing Lumber Composite Price for the period 2000–2017 is also shown in Fig. 2.2. Since 2010, the

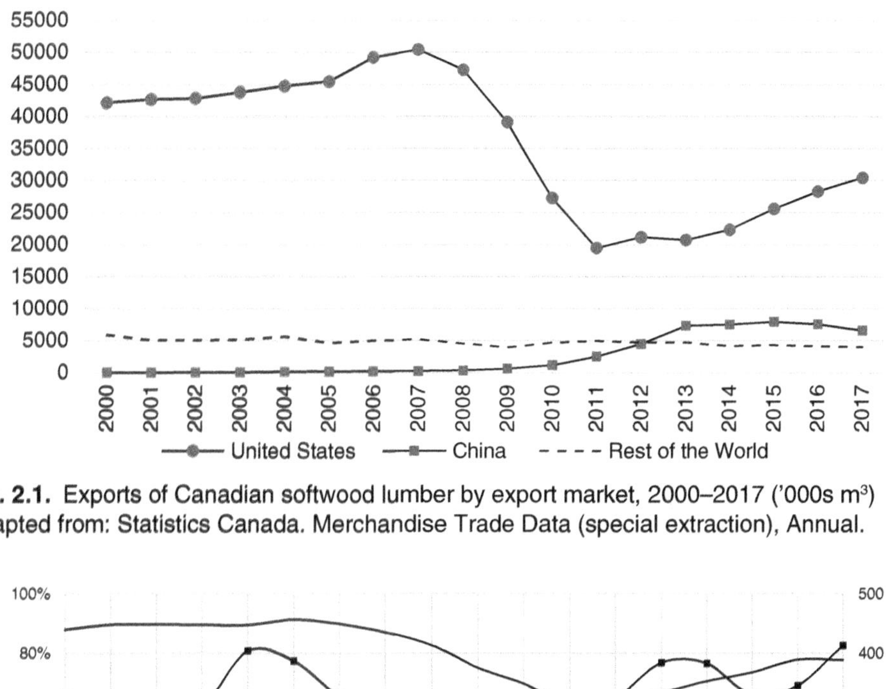

Fig. 2.1. Exports of Canadian softwood lumber by export market, 2000–2017 ('000s m^3) Adapted from: Statistics Canada. Merchandise Trade Data (special extraction), Annual.

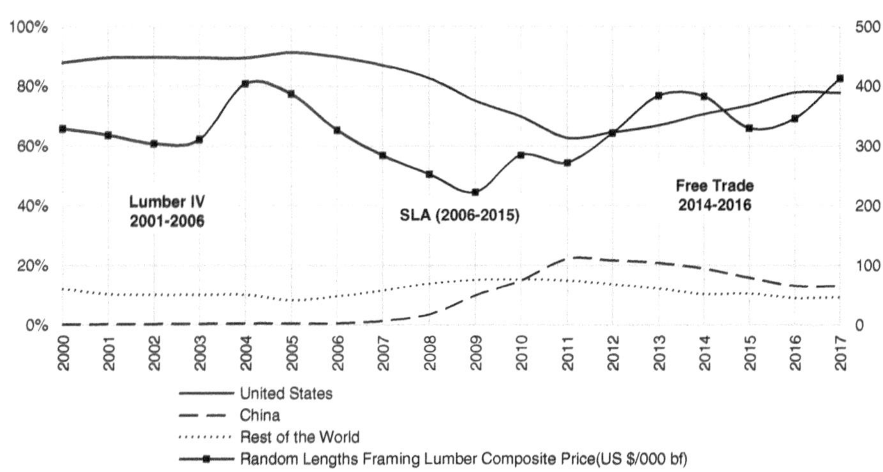

Fig. 2.2. Exports of Canadian softwood lumber by market (%) (left axis) and lumber prices ($/mbf) (right axis), 2000–2017. Adapted from: Statistics Canada (special extraction) and Random Lengths.

monthly price exceeded \$315/mbf (thousand board feet) (\$133.50/m³) on many occasions, leading to a reduction in the export tax. In December 2012, the average price went above \$355/mbf (\$150.40/m³), which enabled Canadian lumber producers to ship tax-free to the U.S. Price volatility continued until 2016, with Canadian lumber producers experiencing long periods where no export tax had to be imposed.

Finally, historical data indicate that British Columbia is by far the largest producer (Fig. 2.3) and exporter (Fig. 2.4) of softwood lumber compared with other regions in Canada. Lumber exports by region throughout the current century are provided in Fig. 2.4. British Columbia (BC) is by far the largest

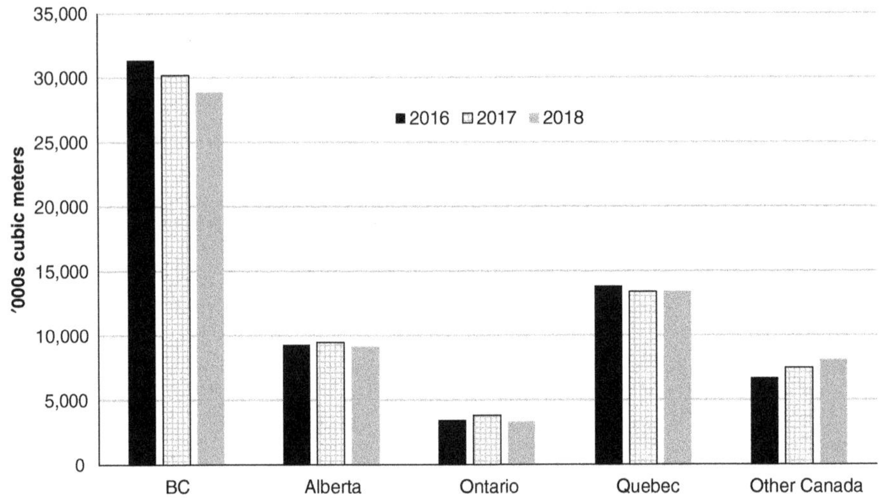

Fig. 2.3. Softwood lumber production in Canada by Province, 2016–2019. Adapted from: Statistics Canada. Table 16-10-0045-01 (https://www150.statcan.gc.ca/n1/tbl/csv/16100045-eng.zip) (accessed September 4, 2020).

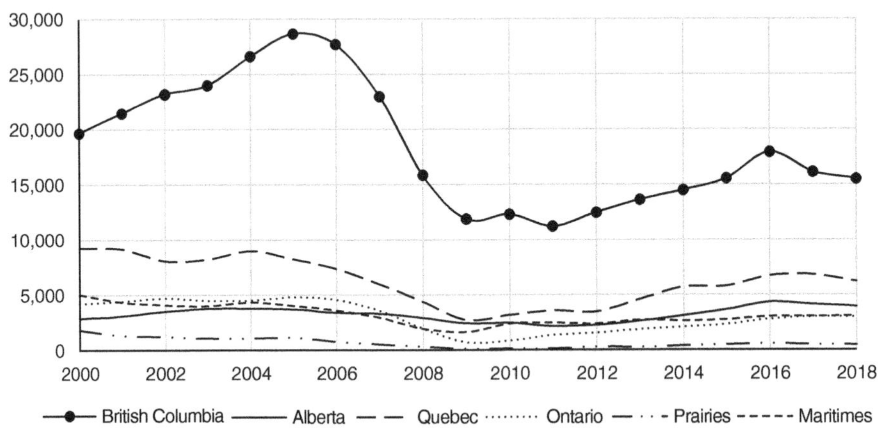

Fig. 2.4. Exports of Softwood Lumber by Canadian Region, 2000–2018 ('000s m³. Adapted from: Statistics Canada. Merchandise trade data (special extraction), annual.

producer and exporter of softwood lumber, followed by Quebec, and, since 2014, by Alberta. Overall, British Columbia is considered to be the major exporter to the U.S., with the province's share of total exports between 48% and 61% during 2000–2018. But it is also clear from the figure that BC is harmed to a much greater extent than other provinces when there is a downturn in the U.S. economy. In 2018, BC exports of softwood lumber reached 24.0 million m^3, with more than 55% shipped to the United States.

2.1 Canadian Forest Tenure Arrangements: A Brief Overview

Canadian forestlands are predominantly owned by provinces, with a variety of tenure arrangements having evolved to facilitate exploitation of forest resources. Long-term tenure agreements or leases extend from 5 to 25 years, and these require the tenure holder to perform duties related to land management as opposed to only harvesting timber. Thus, they are often referred to as land-based tenures. Contrarily, short-term tenures generally grant access to a certain volume of timber that the rights holder can harvest from a particular parcel of land within a given time frame (usually less than 2 years). These are referred to as volume-based tenures that do not usually require the tenure holder to manage the land, although harvest and post-harvest treatments, such as reforestation, are often required.

Provinces manage their forest resources to achieve a variety of policy objectives, including economic (often rural) development (viz. community stability), employment, environmental, revenue, and other objectives. For example, integrated forest companies were granted long-term, land-based tenures over public land as an incentive to establish a processing facility; indeed, construction of a pulp mill or sawmill is often a pre-condition for accessing timber from public lands, a requirement known as **appurtenancy**. In many provinces, establishment of a mill is still a prerequisite for a private company to acquire a long-term, land-based forest tenure. However, BC did away with appurtenancy in 2003, although the government has indicated a willingness to reinstate this policy to protect rural employment (*Times Colonist*, 2018). Other tenure forms are likewise an outgrowth of social policy objectives that differ from those of private forest companies; examples include volume-based tenures that target small logging firms or ones specifically designed to encourage investments in the forest industry by indigenous people.

Tenure arrangements generally place minimum and maximum constraints on the volume of timber that can be harvested over a planning horizon of 1–5 years, while requiring firms to meet sustainable harvest and/or forest management requirements. That is, log harvests are not generally responsive to changes in lumber prices, with too many logs harvested when markets are down and too few when lumber prices are high. In setting stumpage fees, the government is slow to respond to market conditions. Meanwhile, the supply of logs is highly inelastic because harvest levels are planned several years in advance. Thus, changes in stumpage fees are about which party (government or company) captures the resource rent, but harvests are less affected by this decision. Because log supply is inelastic, the supply of lumber is also inelastic, which might explain a predilection for softwood lumber agreements to rely on quotas rather than export taxes or import duties. Of course, Canadian producers will, despite risks of spoilage (e.g. fire), rely on storage of logs and/or lumber to smooth market fluctuations, although this aspect has not been examined in the context of the SWL disputes.

More information about forest tenure systems can be found in Pearse (1990, pp. 172–193) and Marchak *et al.* (1999); many other authors have also discussed the benefits and shortcomings of Canada's forest tenure system. Whether Canada's tenure system can be changed to satisfy softwood lumber producers in the United States is another issue: the Canadian public is generally opposed to the sale of public forestlands to private owners – the status quo in forestland ownership will be retained. As noted in the next section, efforts to change the tenure as opposed to ownership arrangements might satisfy U.S. producers, but this could well be an uphill battle.

2.2 Recent History of the Canada–U.S. Softwood Lumber Dispute

The importance of softwood lumber trade between Canada and the U.S. has been emphasized by Uhler (1991), Delcourt (1995), van Kooten (2002), Berck (2005), and Biggs *et al.* (2006), among others. Although it has a long history, the most recent version of the U.S.–Canada softwood lumber dispute began in 1982 (Cashore, 1997; Marchak *et al.*, 1999). At the heart of this dispute is the claim by U.S. lumber producers that Canadian companies have a comparative advantage because of Canada's timber pricing policies, which constitute a subsidy to processors. Since 1982, this has resulted in five major disputes and three trade agreements. In Canada, about 94% of forestlands are owned and administrated by a provincial government and, to a lesser extent, the federal government, with the remaining 6% privately owned (NRC, 2017). The provinces own 90% and the federal government owns 2% of public lands, known as crown lands. In contrast, in the U.S., 42% of the forestland is owned by the government and 58% is privately held. Therefore, Canada's lumber industry relies primarily on timber logged on public lands, while U.S. lumber producers rely on timber from private lands (Hoover and Fergusson, 2018).

In an attempt to make timber pricing more transparent, BC developed open auctions for some 10 to 13 million cubic meters of timber, or between about one-third and four-fifths of the province's total annual harvest (BC Ministry of Forest, Land, Natural Resource Operations and Rural Development, 2017). The stumpage fees from these auctions are used to inform the prices that the government then sets on timber harvested under longer-term land- and volume-based tenures. However, this seems not to have satisfied U.S. lumber producers, as evidenced by the most recent SWL dispute. To gain insight into the current state of affairs, in the next subsections we summarize the history of the Canada–U.S. softwood lumber dispute and how it was resolved at various times since 1982 (Gorte, 2001; Charron, 2005; Gorte and Grimmett, 2006; Kotyk, 2017; Hoover and Fergusson, 2018). The discussion is instructive as we highlight five lumber disputes and the ways in which they were resolved.

Lumber I (1982–1983)

In late 1982, the U.S. Coalition for Fair Canadian Lumber Imports (the Coalition) filed a preliminary countervailing duty petition with U.S. Department of Commerce (DOC) alleging that provincial stumpage fee systems and Canada's methods of managing public forestlands harmed the U.S. lumber industry. The resulting investigation, completed in May 1983 by the International Trade Administration (ITA) of the United States, found no evidence for the imposition of a countervailing duty (CVD), arguing that the stumpage programs were not targeted to any specific industry or enterprise (Biggs *et al.*, 2006).

Lumber II (1986–1991)

The U.S. Coalition filed a new petition with the DOC in 1986 and this time the DOC altered its calculations to consider the intrinsic or non-market value of the standing timber as part of the costs to Canadian governments. The intrinsic value of the forest would be lost when a forest was logged. The DOC ruled that the stumpage programs created a countervailable subsidy, and so imposed a CVD of 15%. Before the imposition of the subsidy, Canada and the U.S. agreed to a Memorandum of Understanding (MOU). One of the terms was that Canada would impose a levy of 15% on all lumber exported to the U.S., beginning in 1987. The 15% export tax on Canadian lumber was to be in place until Canadian provinces raised their stumpage fees. "By 1991, British Columbia, Quebec and most Atlantic provinces had made sufficient changes to their stumpage systems that export charges could be substantially eliminated. Hence, Canada elected to terminate the MOU in September 1991" (Biggs *et al.*, 2006).

Lumber III and the Softwood Lumber Agreement (1992–1994 and 1996–2001)

Following the termination of the MOU, the United States imposed another CVD of 6.5% in mid-1992 on lumber imports from all provinces except those in the Maritimes. This time, Canada appealed both the determination of subsidy and the finding of injury to the binational panel for dispute resolution under Chapter 19 of the Canada–U.S. Free Trade Agreement (Charron, 2005). The panel rejected the DOC's decision, but, invoking mechanisms under the Free Trade Agreement, the U.S. requested that an Extraordinary Challenge Committee be struck to review the panel's decision – the request was subsequently rejected.

In 1996, the two governments signed a second 5-year MOU, the Softwood Lumber Agreement (SLA). Upon signing, the U.S. refunded the duties that had been collected under Lumber III (Charron, 2005). Under the SLA, a quota device was employed that constrained annual lumber exports to the U.S. from British Columbia, Alberta, Ontario, and Quebec to 14.7 billion board feet (BBF), or 34.7 million m^3, until 2001. The agreement included an escalating fee structure on shipments above quota (Biggs *et al.*, 2006; Hoover and Fergusson, 2018). The SLA posed a challenge for Canada's forest producing regions, except the Maritimes, but particularly for the BC Coast as a result of mill closures and layoffs. The agreement was terminated in 2001.

Lumber IV and the Softwood Lumber Agreement (2006–2015)

Another round of disputes started on April 2, 2001 when the U.S. Coalition filed a new petition with the DOC to impose CVDs. This dispute was by far the most complicated to date as the U.S. recognized five federal and 22 provincial programs as sources of subsidization. As a result, the DOC published a final determination of a 19.23% CVD and an anti-dumping (AD) duty of 8.43%. Canada and individual Canadian companies appealed to the North American Free Trade Agreement (NAFTA) and the World Trade Organization (WTO) for the review of various parts of the CVD and AD determination.

During 2004 and 2005, appeals to NAFTA's dispute-resolution panel required the DOC to revise its estimated duties five different times, in each case favoring Canada, thus causing duties to be lowered. The CVD was lowered from 18.8% in the beginning of 2004 to 17.2% by the end of 2004, and the AD rate dropped from 8.4% to 4%, for a total reduction from 27.2% to 21.2%. Further reductions occurred during 2005 – to a CVD of 16.4% and AD of 3.8%, for a total rate of 20.2%. By the beginning of 2006, the combined CVD and AD had dropped even further – from 20.2% to 10.8%.

As a result of a Canadian government appeal to NAFTA in March 2006, the dispute-resolution panel denied the challenge filed by the Coalition for Fair Lumber Imports, which was fighting to keep the CVD in place. On April 26, 2006, Canada and the U.S. announced that they had reached a tentative 7-year agreement, with an option for two additional years. The deal called for the U.S. to revoke softwood duties and for Canada to implement border measures. Export taxes ranging from 5% to 15% and/or quotas on Canadian shipments to the U.S. would kick in when the Random Lengths' Framing Lumber Composite Price fell below $355/mbf. Each province was allowed to choose between two options, either a tax only (option A, 15% tax rate) or a tax combined with a quota (Option B). The options are described in Table 2.1.

Table 2.1. Triggers and available options under the 2006 SLA.

Random Lengths' Framing Lumber Composite Price	Option A: Export Charge	Option B: Export Charge + Quota
Over US$355/mbf	0%	0% + no quota
US$336 to US$355/mbf	5%	2.5% + regional share of 34% of U.S. consumption
US$316 to US$335/mbf	10%	3.0% + regional share of 32% of U.S. consumption
US$315 or under	15%	5.0% + regional share of 30% of U.S. consumption

Adapted from: *Softwood Lumber Agreement, Article VI, Paragraph 2* (http://www.treaty-accord.gc.ca/text-texte.aspx?id=105072&lang=eng) (accessed July 4, 2016)

This second Softwood Lumber Agreement was finally implemented on October 12, 2006. All covered provinces were charged under Option A. In January 2007, Option B was scheduled to be available to provinces that chose to have a smaller tax combined with a quota restriction. The agreement began with plenty of angst, as Canadians feared the agreement would not be implemented as, the day after the SLA was implemented, the U.S. International Trade Commission ruled that Canada should be refunded the full amount of the duties that had previously been imposed. Because the SLA had already been signed, the ruling was considered moot. The SLA employed region-specific, tariff-rate quotas (TRQs) on softwood lumber imports from Canada, but exempted the Atlantic Provinces. A choice between two border measures was made available to non-exempt provinces in 2006, although provinces could switch options every 3 years, with none having done so. Alberta and BC chose Option A, while Saskatchewan, Manitoba, Ontario, and Quebec took Option B.

As noted in section 2.1, the monthly average of the composite price exceeded $315/mbf after 2010, which led to a reduction in the export tax. When the average price went above $355/mbf in December 2012, Canadian lumber producers could export lumber to the U.S. free of the export tax. Price volatility continued until 2014 with Canadian lumber producers experiencing long zero-tax periods by October 2014, followed in 2016 with prices above $355/mbf, which effectively meant free trade.

Under the SLA, the allocation of TRQs among provinces was a source of controversy because there were times when one or more regions failed to export their quota allotment, while another region exceeded its quota limit. This was a particular problem for Alberta, which often exported softwood lumber in excess of its quota, even to the point where it incurred penalties, as indicated in Fig. 2.5. Economic theory predicts that, by reallocating quota toward Alberta, overall wellbeing would increase as inefficient lumber producers (in regions outside Alberta) would be driven out of the market (Johnston and van Kooten, 2017). By permitting quota trade within and across provinces, however, Alberta producers could bid the 'right-to-export' lumber away from less efficient producers in other provinces. This would keep production costs lower than otherwise. However, it might result in some sawmills in other provinces shutting down, but loss of mills was something each province wished to avoid.

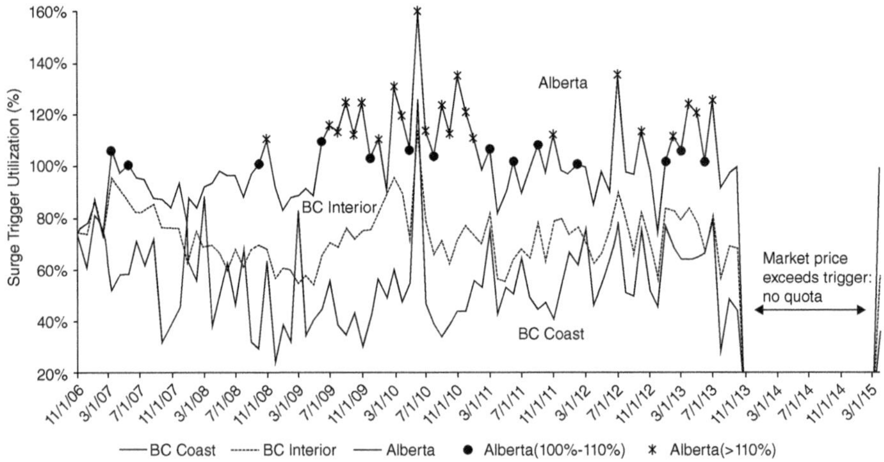

Fig. 2.5. Quota volume utilization (%) under SLA and indications where surge trigger exceeded, select regions: November 2006–April 2015. Adapted from: Foreign Affairs, Trade and Development Canada (2015).

Lumber V

The latest Softwood Lumber Agreement had been struck in September 2006 and later extended to October 2015, followed by a 1-year grace period during which U.S. lumber producers could not initiate action against Canada. The U.S. Coalition initiated a new dispute immediately upon expiration of the 'cooling-off' period. In February 2017, four companies were selected to be investigated in the duty cases, including Canfor, West Fraser, Tolko, and Resolute (see Table 1.2 in Chapter 1, this volume). If an investigation by the U.S. DOC warranted the application of duties, those companies would be assessed company-specific duties and all others would be assessed a weighted average of the duty rates imposed on those four companies. The DOC issued its preliminary determination in the countervailing duty case in April 2017, imposing average duties of 19.88% on Canadian lumber exports to the U.S. The four mandatory respondents in the case were levied the following preliminary CVD rates: West Fraser, 24.12%; Canfor, 20.26%; Tolko, 19.50%; and Resolute, 12.82%. J.D. Irving requested to be a voluntary respondent in the investigation, and the DOC subsequently imposed a much smaller rate of 3.02%. All other Canadian mills were assessed the weighted average of the rates imposed on the five firms that were investigated. The preliminary CVD went into effect on April 28, 2017. West Fraser, Canfor, Tolko, and Resolute were not assessed duties retroactive to January 28, but J.D. Irving and all other Canadian shippers were required to pay retroactive duties.

The U.S. DOC announced its preliminary determination in the anti-dumping duty case on June 26, 2017. The assessed average AD rate was 6.87%, which, combined with a preliminary CVD rate of 19.88%, meant that most Canadian shippers paid total duties of 26.75% on exports to the U.S.

Individual AD rates were as follows: Canfor, 7.72%; Resolute, 4.59%; Tolko, 7.53%; and West Fraser, 6.76%. The DOC also announced that Newfoundland and Labrador, Nova Scotia and Prince Edward Island were excluded from the investigations, and they would not be required to pay duties.

On November 2, 2017, the DOC announced its final determinations in the CVD/AD investigations. Most Canadian shippers thus paid combined duties of 20.83% on lumber shipments to the U.S. Combined CVD/AD rates for the major companies were: Canfor, 22.13%; Resolute, 17.90%; Tolko, 22.07%; West Fraser, 23.76%; and J.D. Irving, 9.92%; the final CVD rate for other companies was set at 14.25% and the AD rate at 6.58% (see Table 2.2). These rates took effect on January 3, 2018, which was the date of the publication of the U.S. DOC's final anti-dumping and countervailing duty orders. However, final duties were not collected until the U.S. International Trade Commission issued an affirmative final injury ruling. Meanwhile, the Canadian government paid some $867 million to support affected workers to reduce the risk of job losses and to stabilize forestry operations, diversify markets and promote innovation in the sector (NRC, 2017).

In April 2019, a WTO panel ruled in favor of the U.S. stance regarding 'zeroing' in the calculation of anti-dumping duties. With zeroing, the foreign domestic price is compared with the U.S. import price adjusted for shipping and handling costs. To determine the anti-dumping duty, a series of foreign domestic prices and matching import prices is required. The AD duty should be an average of the differences, weighted by the amount traded. Upon subtracting the import price from the foreign domestic price, a negative value would indicate lack of evidence of dumping. The U.S. procedure for calculating the AD duty takes the negative differences and treats them as zero, thereby raising the average difference between the foreign domestic price and import price and, thus, the AD duty to be imposed. The panel ruled that this method was acceptable in the softwood lumber case.

Then, in early September, 2019, a NAFTA panel ruled that the claims by American lumber producers that they had been harmed by Canadian producers were flawed. It ruled that the CVD and AD rates imposed on Canadian producers had to be recalculated, and thereby sent the file back to the U.S. International Trade Commission (USITC) for reconsideration.

Table 2.2. Current anti-dumping and countervailing duties by company.

Company	Anti-dumping duty	Countervailing duty	Combined duty
Canfor Corporation	7.28%	13.24%	20.52%
Forest Products Canada Inc.[a]	3.20%	14.70%	17.90%
Tolko Marketing Sales Ltd	7.22%	14.85%	22.07%
West Fraser Mills Ltd	5.57%	17.99%	23.56%
J.D. Irving Ltd	6.04%	3.34%	9.38%
All Others	6.04%	14.19%	20.23%

[a]Formerly Resolute Inc.

Adapted from: Global Affairs Canada, March 3, 2020; at https://www.international.gc.ca/controls-controles/softwood-bois_oeuvre/index.aspx?lang=eng (accessed April 2, 2020).

2.3 Policy Options: Lumber Dispute Remedies

A major consideration in trade negotiations with the U.S. pertains to the use of quota in the allocation of softwood lumber exports, although countervailing and anti-dumping duties or export taxes have also been a feature of negotiations. As noted in the previous section, most resolutions to the various disputes have involved quotas. When it comes to quotas, however, the U.S. lumber coalition essentially seeks to divide Canada by regions or provinces so that the overall quota allocated to Canada is sometimes not met; inefficiencies result because quota cannot be allocated to the regions with the most efficient lumber producers. There potentially exist large benefits for Canadian regions to cooperate in the allocation of quota so that lumber produced for the U.S. market is produced at lowest cost (van Kooten, 2002). The distribution of quota across provinces and among sawmills is an issue that needs to be examined from an economic standpoint. For example, it may be efficient to allocate (grandfather) quota on the basis of a company's historic production or historic sales to the U.S. and then allow trade in quota. This will avoid future situations where some provinces pay over-quota fees while others are unable at the same time to reach their quota limits, as was the case under the 2006 SLA (which resolved Lumber IV).

Lumber is not a homogeneous product and engineered wood products that replace concrete and steel in construction also need to be considered. Future trade models explicitly need to include a variety of lumber and engineered wood products, as some products are substitutes for U.S. lumber, while other Canadian lumber and wood products are complements to U.S. lumber. For example, Mogus *et al.* (2006) found evidence to indicate that Canadian lumber might, at times, be a complement rather than a substitute for U.S. lumber. Sorting this out is extremely important to trade negotiations. Further, it is necessary to determine how policies related to engineered wood products and trade with non-U.S. countries might affect the outcome of SWL trade negotiations.

Four instruments can be used to resolve U.S.–Canada softwood lumber disputes – a straightforward export tax, a straightforward import tariff, a hard quota (that might vary between provinces), and a tariff-rate quota system where both tax rates and quota levels vary according to the outcome of an index that is indirectly related to economic activity and thus demand for lumber. The tariff-rate quota is the most complicated scheme to implement, but it follows the approach permitted under WTO rules; it was also the approach employed in SLA2006. An import duty behaves much like an export tax, except that the revenue accrues to the U.S. instead of Canada. All four instruments have the effect of reducing Canadian exports of softwood lumber to the U.S. below what would occur under free trade, which is their intent. This raises the price received by U.S. lumber producers at the expense of U.S. consumers and Canadian producers (although Canadian consumers would benefit, as prices in Canada would be lower). However, each instrument creates a form of monopoly rent, which we refer to as a quota rent in the cases where a quota instrument is employed. Since U.S. lumber producers appear unconcerned about this rent and are really unable to capture it, the dispute from the Canadian side really concerns the potential size of the rent and how it is allocated or dissipated.

2.4 Modeling Approaches

Our trade modeling approach builds on previous trade models by Mogus *et al.* (2006), Abbott *et al.* (2009), and Johnston and van Kooten (2017). The latter authors explicitly model a TRQ, which is illustrated here with the aid of Fig. 2.6. The U.S. (excess) demand function represents the difference between U.S. domestic demand and supply for prices below the U.S. autarkic price (see below). This is the effective demand for Canadian softwood lumber exports. The ad valorem tax (tariff) for within-quota sales is given by t_0, while the over-quota tax rate is t_1. In the figure, there are no lumber sales beyond the quota and the market clearing price at the quota level is *P*. The tax revenue is given by the lighter shaded area while the quota windfall accruing to Canadian producers is given by the darker shaded area. Together the two shaded areas equal the quota rent. Because we investigate the impact of an ad valorem tax on lumber, the tax revenue and quota rent depend on the prices and quantities, which are determined simultaneously in the model.

One approach to analyzing the economics of the Canada–U.S. softwood lumber dispute employs partial-equilibrium trade modeling. Four such approaches are examined:

1. A spatial price equilibrium (SPE) approach that assumes differences between prices in different countries or regions is the result of shipping and handling (S&H) costs plus any tariffs. It is discussed further below and in Chapters 3, 4, and 5.

2. An alternative SPE model is described in Chapter 6 along with various applications of the model. It assumes that all countries and regions sell or buy lumber from a central clearing house (viz. a Walrasian auctioneer). In contrast to the first approach, where bilateral trade flows are directly modeled, this approach

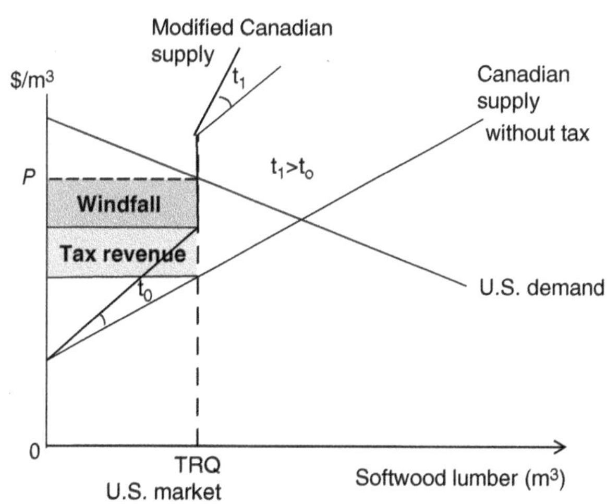

Fig. 2.6. Tariff-rate quota in Canada–U.S. softwood lumber trade: t_0 is the tax up to the quota level, t_1 is the over-quota tax rate. (Authors' own figure.)

can only determine bilateral trade flows in an ad hoc manner after markets are cleared.

3. While the foregoing approaches employ constrained optimization (mathematical programming) to determine trade flows, a gravity model uses econometric analysis to estimate factors that determine trade flows among regions. The attraction or gravity parameter in these models is generally countries' gross domestic product (GDP). The gravity model is discussed in Chapter 7.

4. Finally, a Global Vector Autoregressive (GVAR) model is described and estimated in Chapter 8 (this volume).

In the remainder of this chapter, we examine the benefits of trade – how countries can increase their wellbeing through trade. To do so, we employ a simple, two-country trade model along with the concepts of excess supply and excess demand. We begin in what follows by illustrating how to find the regional excess supply and demand functions that are used in trade modeling. It is in international markets where excess demand and supply functions are used to determine prices that can then be tracked back to domestic/local markets. Further, areas under the excess demand and supply functions represent the gains or losses accruing to importing and exporting countries, respectively. A diagrammatic explanation of spatial price equilibrium trade models, and excess supply (ES) and excess demand (ED) functions, can be found in Just *et al.* (2004), Schmitz *et al.* (2020) and, in the context of forestry, van Kooten and Folmer (2004, pp. 409–421). That literature is summarized in what follows.

2.4.1 Autarky and the excess supply and demand functions

Consider Fig. 2.7, where the domestic supply of and demand for lumber are denoted *S* and *D*, respectively. The equilibrium price and quantity for lumber under autarky (no trade) are p_1 and q_1. Suppose the price rises above p_1 for whatever reason. If it were to rise to p_2, the country will produce q_2 but only consume q_3. In other words, the country will produce $q_2 - q_3$ more lumber than it would consume at the given price p_2. The quantity available for export for any price above the autarkic price of p_1 is given by the horizontal difference between the quantity produced and the quantity demanded for a given price – these differences constitute the excess supply (ES) function. Thus, at the price p_2, the excess supply of lumber is q_4, which is exactly equal to $q_2 - q_3$. The area above the ES curve below a given price is a measure of the gains from trade. This gain equals area *W*, which is exactly equal to area *X*, and is the excess of producer surplus gain over the consumer surplus loss as a result of moving from the autarkic equilibrium price p_1 to p_2.

Similarly, the excess demand curve is represented in Fig. 2.7 by price deviations below the autarkic price. Suppose, as a result of free trade, price falls from p_1 to p_3. Consumers desire quantity q_2 while producers only supply q_3, so imports of amount $q_2 - q_3$ occur. The excess demand schedule ED is derived

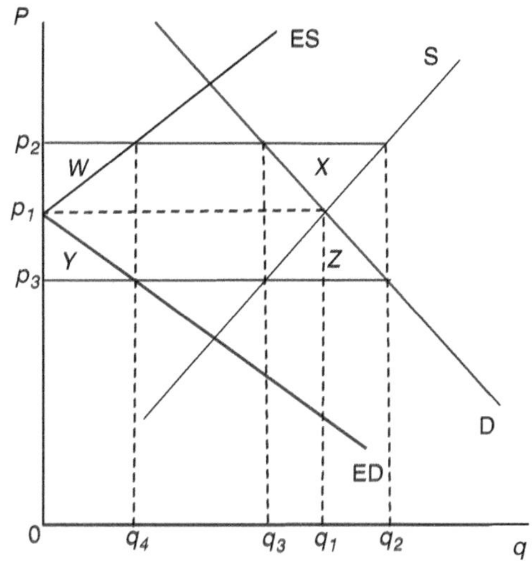

Fig. 2.7. Excess supply and excess demand. (Authors' own figure.)

by horizontally subtracting quantities supplied along the supply curve from quantities demanded along the demand curve. For example, if the price were to fall below p_1 to p_3, the country would import q_4, which is exactly equal to $q_2 - q_3$. The area below the ED curve bounded by a given price is a measure of the gains from trade. This area equals Y, which also equals area Z that, in turn, is the excess of consumer surplus gain over the producer surplus loss as a result of moving from the autarky equilibrium price p_1 to p_3.

The ES and ED schedules can be derived mathematically. Suppose the (inverse) demand and supply curves in Fig. 2.7 are linear:

$$p^D = \alpha - \beta q, \ \ \alpha, \beta \geq 0, \tag{2.1}$$

and

$$p^S = a + bq, \ a, b \geq 0. \tag{2.2}$$

The excess demand and supply curves in the figure are then given by:

$$\text{ED} = \gamma - \delta q, \text{ with } \gamma = \frac{a\beta + b\alpha}{\beta + b} \geq 0 \text{ and } \delta = \frac{b\beta}{\beta + b} \geq 0. \tag{2.3}$$

$$\text{ES} = \gamma + m q, \text{ with } \gamma = \frac{a\beta + b\alpha}{\beta + b} \geq 0 \text{ and } m = \frac{b\beta}{\beta + b} \geq 0 \tag{2.4}$$

Notice that γ (= p_1) is the equilibrium domestic price in autarky, such that in the absence of shipping and handling (S&H) costs, the excess supply and demand curves start at the same point on the vertical (price) axis. Further, the absolute slopes of the ED and ES curves are identical (at least to the choke prices), although ED slopes down and ES slopes up.

For grammatical convenience, consider first lumber trade between only two countries or regions, A and B. The two-country example offers an excellent way to illustrate how spatial price equilibrium trade models can be used to analyze policy. A numerical mathematical programming model would be required

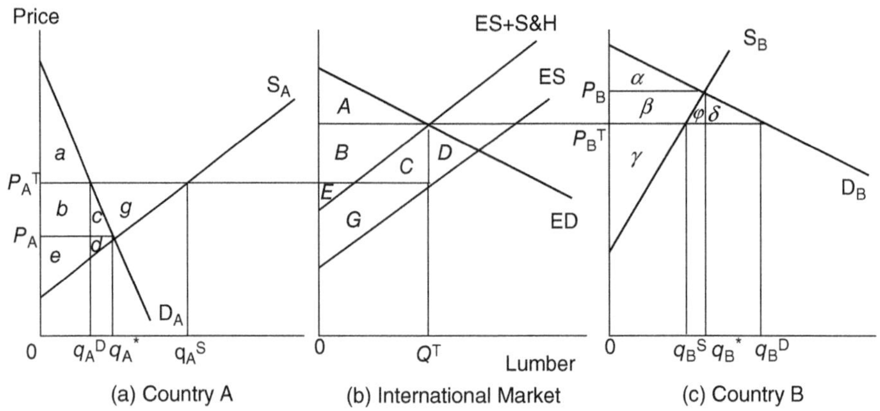

Fig. 2.8. Two-country model of lumber trade. (Authors' own figure.)

to model the real world as it is characterized by bilateral trade among many countries or regions. The two-country spatial price equilibrium lumber trade model is illustrated in Fig. 2.8.

In the figure, the domestic demand functions in countries A and B are given by D_A and D_B, respectively, while the domestic supply functions for lumber are given by S_A and S_B. Under autarky, an amount q_A^* will be supplied and consumed in country A, at a domestic price of P_A (see Fig. 2.8a); in country B, the autarkic price and quantity are P_B and q_B^* (Fig. 2.8c). Note that, for trade to take place, the difference between autarkic prices must exceed the shipping and handling (S&H) costs of transporting the good from one market to the other; that is, $|P_A - P_B| > t$, where t refers to the S&H costs.

Economic wellbeing or welfare is always determined as the sum of surpluses (Just *et al.*, 2004); that is, the wellbeing of citizens in each country is determined by the sum of the benefits they receive as consumers (consumer surplus) and as producers (quasi-rent), plus any surplus that can be attributed to the country's natural resources. In the absence of trade, the consumer surplus associated with the lumber market in country A is given by area (a+b+c) in Fig. 2.8(a), and area α in Fig. 2.8(c) for country B. The producer surplus in the absence of trade is measured by area (e+d) for country A and (γ+β) for country B. Total economic wellbeing is the sum of producer and consumer surpluses, and is simply given by the area between the demand and supply curves. For country B, total surplus in the absence of trade is given by area (α+γ+β), while it is (a+b+c+d+e) for country A.

2.4.2 Gains from trade

To demonstrate that trade improves the wellbeing of both countries, it is necessary to show that the total surplus in each region increases as a result of trade. In the absence of trade, the price in country B exceeds that in country A (Fig. 2.8). With trade, the price in country B falls from P_B to P_B^T, while country A's price rises from P_A to P_A^T. Consumers in country B gain as a result of the price decrease; consumption rises from q_B* to q_B^D and consumer surplus increases from area α to area $(\alpha+\beta+\phi+\delta)$. However, producers in country B face a lower price, namely, $P_B^T<P_B$ in panel (c), causing them to reduce production from q_B* to q_B^S. An amount $q_B^D-q_B^S$ is purchased from country A, while producer surplus falls from $(\gamma+\beta)$ to γ. The overall wellbeing of country B increases by area $(\phi+\delta)$, with consumers (home builders, furniture makers, etc.) as the main beneficiaries from trade.

The situation in country A mirrors that of B. The rise in country A's prices causes consumers to purchase less lumber (from q_A* to q_A^D) and reduces their overall consumer surplus by area $(b+c)$. Producers in country A now receive a higher price and ramp up their production of lumber from q_A* to q_A^S, leading to an increase in producer surplus of $(b+c+g)$ in the process. The overall wellbeing of country A increases by area g, with producers (manufacturers of lumber) the main beneficiaries from trade.

The main results can be summarized in the international market, Fig. 2.8(b). The amount traded between A and B is $Q^T = q_A^S-q_A^D = q_B^D - q_B^S$. The net gain to country B is area A, which is equal to area $(\phi+\delta)$ in panel (c); the net gain accrues to consumers in country B, and is therefore measured under the excess demand curve. Meanwhile, the net gain to country A is the area $(E+G)$ in panel (b), which is equal to area g in panel (a); this gain accrues to the producers of lumber. Note that S&H costs equal $(B+C) = (G+C)$ in Fig. 2.8(b), which implies that area B equals area G.

Given this background, the next step is to investigate the economic theory underlying a more complicated trade model, one that considers vertical and horizontal chains and more than two regions. The questions that need to be addressed are as follows. If a policy affects one market, what welfare measures do we need to consider in inter-related markets? How does a policy in one market affect markets in the vertical chain, and markets for complements and substitutes that are considered horizontal to the market in question? These questions are addressed in Chapter 3.

References

Abbott, B., Stennes, B. and van Kooten, G.C. (2009) Mountain pine beetle, global markets and the British Columbia forest economy. *Canadian Journal of Forest Research* 39(7), 1313–1321.

BC Ministry of Forest, Land, Natural Resource Operations and Rural Development (2017) BC timber sales business plan 2017/18 to 2019/20. Available at: https://www2.gov.bc.ca/assets/gov/farming-natural-resources-and-industry/forestry/bc-timber-sales/business-plans-performance-reports/bp-201718-201920.pdf (accessed November 7, 2019).

Berck, P. (2005) Contested trade in logs and lumber. In: Schmitz, A., Moss, C.B., Schmitz, T.G. and Koo, W.W. (eds) *International Agricultural Trade Disputes: Case Studies in North America.* University of Calgary Press, Calgary, Canada.

Biggs, J., Laaksonen-Craig, S., Niquidet, K. and van Kooten, G.C. (2006) Resolving Canada–U.S. trade disputes in agriculture and forestry: lessons from lumber. *Canadian Public Policy,* XXXII(June), 143–155.

Cashore, B. (1997) Flights of the phoenix: explaining the durability of the Canada–U.S. softwood lumber dispute. *Canadian-American Public Policy* 32, December, 1.

Charron, M. (2005) Lumber I to IV, history of Canada–U.S. softwood lumber dispute. Library of Parliament TIPS-134E, Ottawa, Canada.

Delcourt, G.V. (1995) Global softwood lumber trade: a spatial equilibrium model. MSc thesis. University of British Columbia, Vancouver, Canada.

Foreign Affairs, Trade and Development Canada (2015) Special data request from the Softwood Lumber Controls Division, 15 May, Ottawa, Canada.

Gorte, R.W. (2001) Softwood lumber imports from Canada: history and analysis of the dispute. CRS Report RL30826. 2 February. Congressional Research Service, Washington, DC. Available at: https://www.everycrsreport.com/reports/RL30826.html and https://file.wikileaks.org/file/crs/RL30826.pdf (accessed July 20, 2020).

Gorte, R.W. and Grimmett, J.J. (2006) Softwood lumber imports from Canada: issues and events. CRS Report RL33752. 15 December. Congressional Research Service, Washington, DC. Available at: https://file.wikileaks.org/file/crs/RL33752.pdf (accessed November 22, 2018).

Hoover, K, and Fergusson, I.F. (2018) Softwood lumber imports from Canada: current issues. CRS Report R42789. 12 April. U.S. Congressional Research Service, Washington, DC. Available at: https://fas.org/sgp/crs/misc/R42789.pdf (accessed November 22, 2018).

Johnston, C.M.T. and van Kooten, G.C. (2017) Impact of inefficient quota allocation under the Canada–U.S. softwood lumber dispute: a calibrated mixed complementarity approach. *Forest Policy and Economics* 74, 71–80.

Just, R.E., Hueth, D.L. and Schmitz, A. (2004) *The Welfare Economics of Public Policy. A Practical Approach to Project and Policy Evaluation*. Edward Elgar, Cheltenham, UK.

Kotyk, A. (2017) Tensions over time: a primer on the Canada–U.S. softwood lumber dispute, *Globe and Mail*, 9. Available at: https://www.theglobeandmail.com/news/british-columbia/tension-over-timber-a-primer-on-canada-us-softwood-lumber-trade/article34942091/ (accessed November 22, 2018).

Marchak, P.M., Aycock, S.L. and Herbert, D.M. (1999) *Falldown. Forest Policy in British Columbia*. David Suzuki Foundation and Ecotrust Canada, Vancouver, Canada.

Mogus, A., Stennes, B. and van Kooten, G.C. (2006) Canada–U.S. softwood lumber trade revisited: examining the role of substitution bias in the context of a spatial price equilibrium framework. *Forest Science* 52(4), 411–421.

NRC (2017) The state of Canada's forests: 2017, annual report. Natural Resources Canada, Ottawa, Canada. Available at: https://cfs.nrcan.gc.ca/publications/download-pdf/38871 (accessed April 2, 2020).

Natural Resources Canada (NRC) (2018) How does the forest industry contribute to Canada's economy? Available at: https://www.nrcan.gc.ca/forests/report/ economy/16517 (accessed September 27, 2019).

Pearse, P.H. (1990) *Introduction to Forestry Economics*. University of British Columbia Press, Vancouver, Canada.

Schmitz, A., Moss, C., Schmitz, T.G., van Kooten, G.C. and Schmitz, H.C. (2020) *Agricultural Policy, Agribusiness and Rent Seeking Behavior*, 3rd edn. University of Toronto Press, Toronto, Canada (in press).

Times Colonist (2018) 'NDP to Restore Timber Harvest Rules, Horgan says', 26 February. Available at: https://www.timescolonist.com/business/ndp-to-restore-timber-harvest-rules-horgan-says-1.23184965 (accessed November 22, 2018).

Uhler, R.S. (1991) Canadian public timber pricing and the great subsidy debate. In: Uhler, R.S. (ed.) *Canada–United States Trade in Forest Products*. University of British Columbia Press, Vancouver, Canada.

van Kooten, G.C. (2002) Economic analysis of the Canada–United States softwood lumber dispute: playing the quota game. *Forest Science* 48, 712–721.

van Kooten, G.C. and Folmer, H. (2004) *Land and Forest Economics*. Edward Elgar, Cheltenham, UK.

3 Spatial Price Equilibrium Trade Modelling: Theory

G. Cornelis van Kooten[1] and Craig M.T. Johnston[2]

[1]*University of Victoria, Victoria, BC Canada;* [2] *Bank of Canada, Ottawa, ON Canada*

Abstract

The theory of measuring economic welfare across horizontal and vertical chains is provided in this chapter. As demonstrated, it underlies the spatial price equilibrium (SPE) trade models. The vertical and horizontal chains involved in a particular trade model, referred to as the REPA model, are described as an illustration.

In the previous chapter, we used the concepts of excess supply and excess demand to demonstrate that two countries could gain from trade. The analysis was limited to the case where the countries traded a single commodity and that no other markets were impacted by the exchange – with the exception of the lumber market, no other goods or services were assumed to be impacted as a result of lumber trade. If trade resulted in lower prices of lumber, say, the price of housing was unaffected; likewise, the prices of other construction materials, whether substitutes such as concrete or complements such as wood panels, or those of any other commodity outside of construction, remain unaffected. These types of assumptions have implications for the development of trade models. What assumptions are required in order to ignore the effects of changes in the lumber market on other markets in the vertical or horizontal chain? When do we need to measure welfare in vertical or horizontal markets? These questions are addressed in this chapter. Then, in Chapter 4 (this volume), we examine several versions of the latest REPA forest trade model and some modeling results.

3.1 Vertical and Horizontal Chains in Forest Trade Modeling

In Chapter 2, section 2.3, the impacts of changes in the lumber market on vertically and horizontally related markets were ignored. In many situations

10.1079/9781789248234.0003

this is not realistic. As an illustration, suppose the government imposes a quota on production of softwood lumber. This will inevitably impact local lumber prices, causing a wedge between the price domestic consumers pay (higher than under free trade) and the marginal cost of producing the lumber (lower than under free trade). The profits earned by sawmills will increase – a reduction in producer surplus will be more than offset by the policy-created quota rent. Upstream forestland owners (including government) and those engaged in logging will see a reduction in demand and, as a consequence, a lower price for logs. The higher price of lumber will be passed down the marketing chain, increasing the cost of building homes and making furniture and thereby incentivizing local firms to substitute concrete, aluminum and steel for lumber in construction.

Suppose instead that the quota is imposed only on exports, leaving domestic markets to clear at the price where domestic supply and demand intersect. The domestic price will then fall, but the price of lumber in the foreign, importing country will rise, more logs will be harvested in the foreign country (but not enough to reduce price to where it was when trade was unimpeded), concrete and other building materials will substitute for lumber in construction, and so on. To determine the ultimate cost to society, and the income transfers that take place, it is necessary to track the changes in welfare across all of the affected vertical and horizontal markets. Despite the fact that the linkages could be complex, they should be considered when evaluating the impact of a policy affecting the forest products industry.

In this section, we establish a framework for evaluating the welfare effects of price changes in vertically and horizontally related markets. The motivating policy is a quota imposed on lumber exports in lieu of an export tax or import tariff by one's trading partner. The analysis begins by considering a marketing chain for logs used in the production of lumber, ultimately consumed by construction, furniture, and other lumber users. Then, in the next section, the analysis is extended to consider multiple users of logs – not just lumber, but producers of plywood, oriented strand board (OSB), medium-density fiberboard (MDF), pulp, wood pellets, and so on. The fiber from the initial harvest of logs is competitively distributed to multiple processors through horizontally integrated markets. These complex vertical and horizontal relationships suggest a framework for establishing a global trade model of forest products that is rooted in economic theory.

3.1.1 Vertical chain integration

To motivate the discussion of the underlying theory and assumptions that enable the integration of vertically connected markets in a trade model, we begin by distinguishing between two types of supply and demand curves. An **intermediate** supply or demand curve refers to a relation between price and quantity that does not take into account the effect of changes in the prices of the commodity in question on **related** goods or services. That is, it does not take into account the rebound effect that the policies in one market have on the prices in related markets that, in turn, affect the demand or supply in the original market. The **general equilibrium** supply and demand functions take these rebound effects into account. It is appropriate to measure the consumer and producer surpluses as areas under the general equilibrium demand and supply curves, respectively (Just *et al.*, 2004).

To motivate the notion of general equilibrium demand and supply, we employ Fig. 3.1 and assume that the price of lumber falls from P^0_{lum} to P^1_{lum}. To derive the competitive supply curve for lumber that accounts for input prices (denoted r), consider a competitive lumber industry that uses logs as inputs. We use r to denote input prices and P to denote prices in downstream markets.

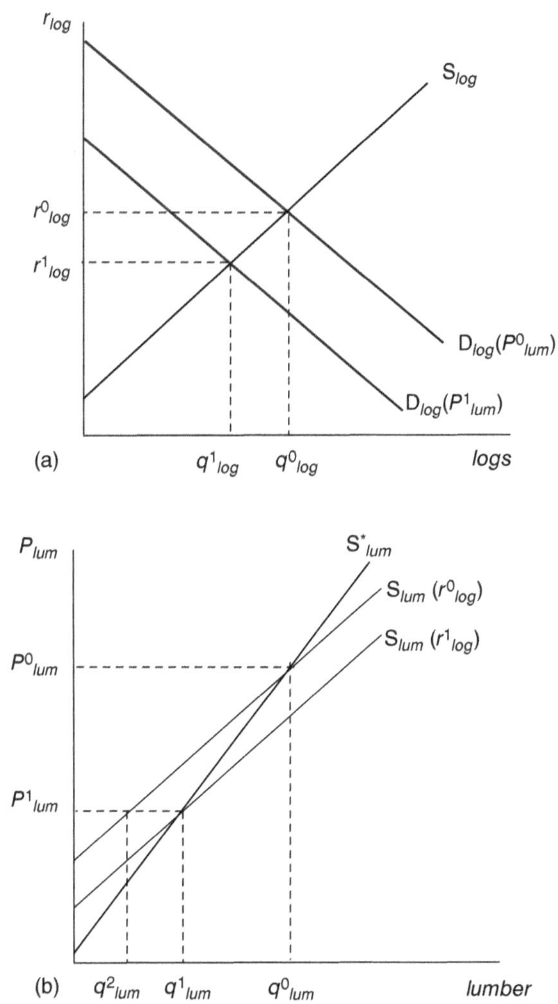

Fig. 3.1. Derivation of the general equilibrium competitive supply curve for lumber in a vertically integrated market chain. (Author's own figure.)

Thus, r is used in the case of logs and markets upstream of logs, and P for lumber and markets downstream. The log market is assumed to have a perfectly elastic supply for inputs (fixed input prices); that is, fluctuations in inputs used to produce logs are assumed not to affect the prices of logging equipment, trucks, fuel, workers, etc. (This is illustrated further in the discussion concerning Fig. 3.2.) Contrarily, the downstream lumber market depicted in Fig. 3.1(b) is characterized by an upward sloping, intermediate supply curve $S_{lum}(r^0_{log})$, and an initial input price of r^0_{log} and output price of P^0_{lum}. The demand for logs in Fig. 3.1(a) is derived from the downstream manufacturers of lumber, as logs are the single most important input into the production of lumber – the demand for logs by lumber producers is given by the value of the marginal product of logs in the production of lumber, or the marginal physical product of logs in the production of lumber multiplied by the output price of lumber. Given a derived demand of $D_{log}(P^0_{lum})$, the log market has initial price r^0_{log} along its intermediate supply curve S_{log}.

Suppose the output price in the lumber market falls to P^1_{lum} as the result of policy intervention. Initially, manufacturers of lumber adjust their production to q^2_{lum} along their initial intermediate supply curve, as they do not perceive the effects of their decisions on the (related) log market. As a result of the price change in the lumber market, the derived demand for logs falls to $D_{log}(P^1_{lum})$, thereby reducing the price of logs to r^1_{log}. In turn, the lower input price leads to a downward shift from $S_{lum}(r^0_{log})$ to $S_{lum}(r^1_{log})$ in the intermediate supply curve for lumber, giving rise to a new equilibrium output and price combination of q^1_{lum} and P^1_{lum}. One can then derive a general equilibrium supply curve for lumber (denoted S^*_{lum}) by connecting the original and final equilibrium points. The general equilibrium supply function allows for equilibrium adjustments of input use and input price, as output price changes. In Fig. 3.1, S^*_{lum} differs from the intermediate supply curve, $S_{lum}(r^0_{log})$, as the latter only indicates how the lumber market will respond to price fluctuations under the premise that input prices remain fixed. The same approach can be used to derive general equilibrium supply and demand functions in other markets; it is areas underneath or above these functions that constitute costs and benefits – the welfare measures.

Given this background, we can see how one measures the welfare implications of policy intervention in multiple vertically and horizontally connected markets. Consider Fig. 3.2 and suppose a quota of q^R_{lum} (or an equivalent ad valorem tax) is imposed on the producers of lumber. This causes the general equilibrium supply curve for lumber, S^*_{lum}, to become vertical at q^{Rlum} so that S^R is the relevant supply function. Lumber is an input into downstream industries such as (primarily) construction, furniture making, and other activities, where it is assumed that the demand in this market is perfectly elastic, D_{n+1}. As a result of reduced lumber production, the price home builders, furniture makers, and other consumers of lumber must pay rises to P^R_{lum}, while the price lumber producers receive falls to P^2_{lum}. Since the demand curve for lumber is derived from the demand for these downstream products, the reduction in consumer surplus, as given by area (a+b) in panel (a), is equal to the change in quasi-rent in the market for downstream users of lumber – the market denoted Q_{n+1} in panel (b) of Fig. 3.2. The loss in producer surplus measured in the downstream market

Q_{n+1} in panel (b) equals area ($\alpha+\beta$), which is identical to area ($a+b$) in panel (a). It is necessary to measure only one of these changes, say area ($a+b$) in panel (a), and not the equivalent loss of area ($\alpha+\beta$) in the downstream market for users of lumber in panel (b).

Consider the income distributional consequences of the quota on lumber. Area ($b+d$) in Fig. 3.2(a) constitutes the deadweight loss triangle due to the inefficiency resulting from the misallocation of resources that results from the restriction on lumber output. Area ($a+c$) represents the policy-induced quota rent that accrues to lumber producers, although it would accrue to the government if a tax were employed to reduce production of lumber by the same amount. Area a in panel (a) equals area α in panel (b), so b equals β. The former represents the income transfer from downstream lumber users (consumers) to lumber producers (or to government), while β (=b) represents the loss due to the subsequent misallocation of resources in the downstream market.

Finally, recall the assumption that the demand function D_{n+1} in the downstream market is infinitely elastic (horizontal). This assumption is necessary to preclude the need to measure welfare gains and losses in markets farther downstream. If the prices of housing, furniture, etc., were impacted by changes in the price of lumber – that the changes $q^1{}_{n+1} - q^0{}_{n+1}$ in the Q_{n+1} market were sufficiently large – then it would be necessary to consider changes in the welfare of home buyers, purchasers of furniture, and so on. We avoid having to measure this loss in consumer surplus by assuming perfectly elastic demand in the Q_{n+1} market.

Now consider the upstream impacts of the restriction in the lumber market. First, in Fig. 3.3, we examine the link between the lumber market (panel b) and the upstream log market (panel a). It is assumed that all of the logs produced by the logging sector are inputs into lumber production. Area ($c+d$) in the lumber market represents the loss in quasi-rent accruing to lumber producers, where area c represents the producer surplus transferred to the policy-induced rent, which might well accrue to producers if a quota is employed but, in the case

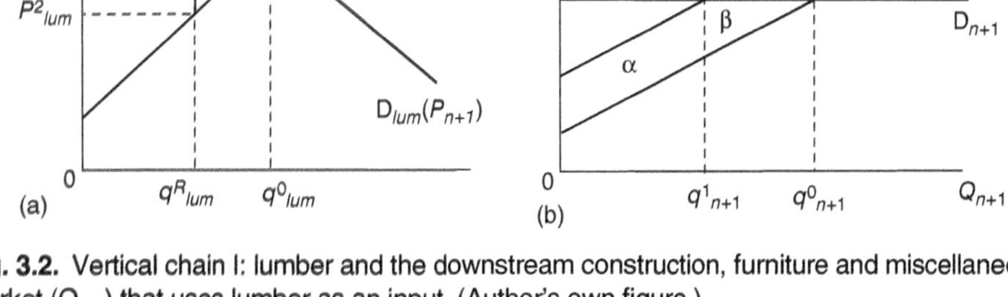

Fig. 3.2. Vertical chain I: lumber and the downstream construction, furniture and miscellaneous market (Q_{n+1}) that uses lumber as an input. (Author's own figure.)

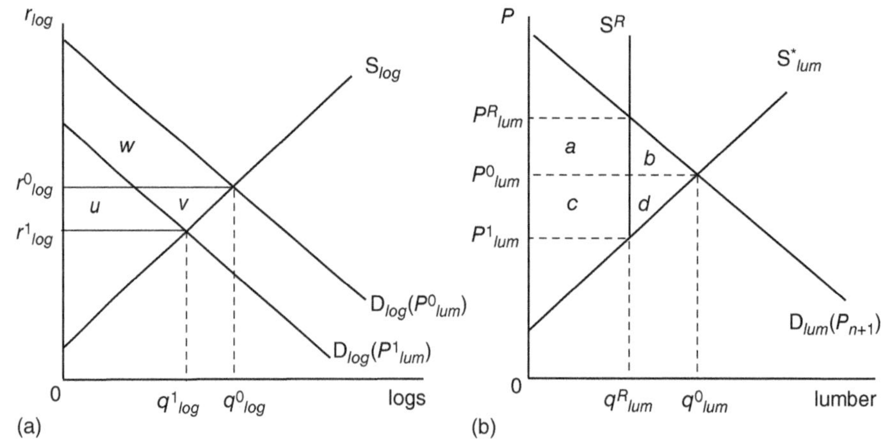

Fig. 3.3. Vertical chain II: integrated log and lumber markets. (Author's own figure.)

of a tax, accrues to government; again area *d* represents a deadweight loss due to the misallocation of resources. Because lumber production has fallen, the derived demand for logs shifts inwards as the effective supply price (marginal cost) of lumber has fallen from P^0_{lum} to P^1_{lum}. Recall that the derived demand for logs is equal to the value of the marginal product of logs in the production of lumber, which equals the price of the lumber multiplied by the marginal physical product of logs in lumber production.

The inward shift in the derived demand for logs results in a loss of consumer surplus in Fig. 3.3(a) equal to area *w* prior to the reduction in the price and quantity of logs sold. To this must be added area *u* which represents a gain in consumer surplus because the price of logs has fallen. The overall loss in consumer surplus is given by area (*w*–*u*) in the log market, which exactly equals area *d* in the lumber market. Notice that the general equilibrium competitive supply curve for lumber S^*_{lum} takes into consideration the effect of the new log price, P^1_{log}, on lumber supply; thus, it is not necessary to have S^*_{lum} shift as a result of this price change as it is inherently incorporated through its derivation (as discussed in conjunction with Fig. 3.1).

Now consider the vertical markets in Fig. 3.4, where the log market is downstream of the input supply market, denoted *n*–1, which provides equipment, trucks, fuel, labor, etc., to logging companies. It is assumed that the input supply schedule S_{n-1} facing market *n*–1 is perfectly elastic so that input prices r_{n-1} are not affected by changes in the demand for such inputs as a result of changes in the harvest of logs. In the log market (Fig. 3.4(b)), the loss in quasi-rent to log producers is equal to area (*u*+*v*), which is equivalent to the change in consumer surplus of area *y* in the upstream market for logging equipment, trucks, fuel, labor, etc. (Fig. 3.4(a)). Because it would be onerous to estimate the demand for inputs into the production of logs, the change in consumer surplus in the input market (panel a) is best measured by the loss in producer surplus in the downstream log market.

The point of the above analysis is this: the welfare measures appropriate for a vertically integrated forest trade model are the consumer surplus, producer

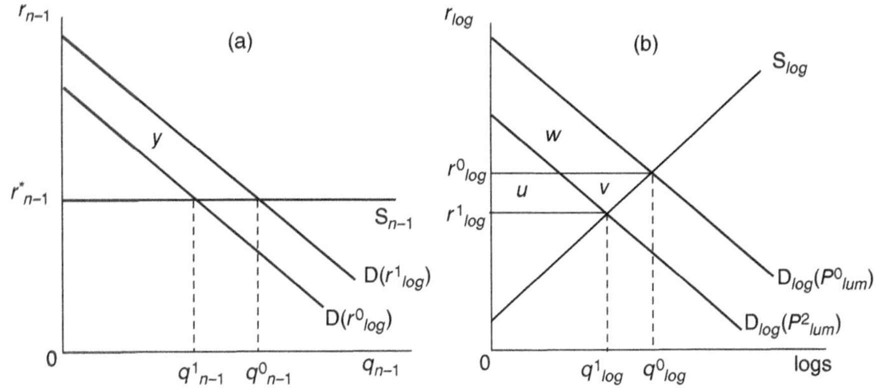

Fig. 3.4. Vertical chain III: integrated log and upstream input supply markets. (Author's own figure.)

surplus (quasi-rent), and a scarcity rent equal to area (a+c) displayed in Fig. 3.2(a) or Fig. 3.3(b). Producers of lumber may capture the scarcity rent if it were created through a quota on production, while if it arose due to an ad valorem tax the government captures it as tax revenue. All of the changes in welfare can be measured in the lumber and log markets, and there is no need to measure welfare in markets upstream from logs or downstream from lumber, because we assume that remaining upstream and downstream markets are characterized by perfectly elastic input supply and output demand, respectively. As noted above, area (*w*–*u*) in the log market is identical to area *d* in the lumber market. Only area *v* in the log market needs to be considered in the cost–benefit (welfare) analysis, but this does not require modeling the log market – information on the elasticities of supply and demand is not needed. Area *v* can be determined as the area of a triangle that is calculated as one-half of the change in the quantity of logs produced multiplied by the change in the price of logs, or $\frac{1}{2} \times \Delta q_{logs} \times \Delta r_{logs}$.

In addition to assumptions of infinite supply elasticity in the upstream markets and infinite demand elasticity in downstream markets, the welfare measures employed are also predicated on the assumption that other wood product markets (markets that are horizontal to lumber) are characterized by a perfectly elastic demand function or that lumber production is the only downstream use of logs. We now consider what happens if this is not the case.

3.1.2 Horizontal chain integration

So far, we have considered the vertical relationship between one input and one output, with additional upstream and downstream markets considered as having respective infinitely elastic supply (fixed input prices) and demand (fixed output prices). Now consider a vertical chain such as that discussed above, but with several outputs from logs and not just lumber; that is, we consider multiple products that use wood fiber from the harvests of timber. The addition of these other wood product markets horizontal to lumber in the marketing chain adds complexity, but also greater reality, to the model.

To add to the challenge of integrating horizontal markets into the discussion, logs are not only an input to downstream processors. With the exception of plywood, wood products rely on residual fiber from sawmilling, logging residues, and whole logs that are chipped because they are unsuited for lumber or plywood. That is, wood fiber might also flow between two or more horizontal markets. For example, the markets for lumber and wood pellets are on the same horizontal market segment (i.e. downstream of logs in the vertical supply chain), because both utilize fiber that originates with the initial log harvest. However, the manufacture of lumber produces, among other things, sawmill residuals that are commonly used in the production of most other wood products, including wood pellets. Wood pellets and lumber are *complements* in production in the sense that by-products from lumber manufacture are used to produce pellets. The more lumber that is produced, the more fiber becomes available to produce wood pellets. Alternatively, two products may be *substitutes* in production if they compete for the same input. Lumber and plywood manufacturers compete for the same industrial roundwood, with logs used to produce lumber (plywood) not available to produce plywood (lumber).

Some products are both complements and substitutes in production. For example, pulp chips are residual to lumber and plywood production, but whole logs can be chipped and used solely to produce pulp. This is true, just as well, for wood chips, OSB, and some other products. One caveat should be noted, however: some logs are not suited to the production of lumber or plywood and might only be worth chipping for pulp purposes or used as a biomass fuel. In practice, the value of logs in lumber is generally much higher than in other uses, so that harvests might not even take place unless the roundwood logs are designated for processing into lumber. Exceptions occur where plantations of fast-growing species such as hybrid poplar have been established to service a biomass power plant or pulp mill.

To demonstrate the importance of the distinction between complements and substitutes in production, consider the expansion of the vertically integrated market structure discussed above. Suppose we have K markets horizontal to lumber, differentiated according to whether they are primarily complements or substitutes to the production of lumber. Let i (= 1, …, k) denote wood products that are joint products (complements) in the production of lumber or plywood, and j (= k+1, …, K) denote wood products that are competitive in (substitutes to) the production of lumber. Further, let $S_i^*(P_i; P_{i-}, P_j, P_{lum})$ and $S_j^*(P_j; P_{j-}, P_i, P_{lum})$ be the respective supply curves for complements and substitutes that incorporate equilibrium adjustments of log inputs and their prices (as indicated by the asterisk), and P_{i-} and P_{j-} denote prices of joint and competitive products, respectively, other than that of the product under consideration. The equilibrium supply functions do shift, however, with changes in the prices of horizontal products.

A change in the price of lumber will lead to changes in relative prices across h horizontal markets. If markets are competitive, an increase in the price of lumber will lead to greater use of fiber in lumber production: (i) logs will be cut more efficiently to produce more lumber (e.g. computers are used to cut logs to obtain the greatest lumber value); (ii) logs will be competed away from other log processors (e.g. by plywood manufacturers); and/or (iii) harvests of commercial roundwood logs will increase. In the first case, the supply of fiber available to complementary products, such as wood pellets and pulp, will decline. In the second case, it is not clear if the amount of residual fiber from plywood manufacture, for example, is more or less than with lumber manufacture.

Consider the third case. Both plywood and lumber production can usually be increased only by harvesting more logs. With increased log harvests, there will be more logging residues and residuals from sawmilling and plywood production – fiber availability for other wood products will increase. At the same time, an increase in the production of lumber/plywood will increase demand for wood products that are complementary to lumber. An example is OSB sheets, which are used in framing houses along with lumber. Given that lumber remains the major driver of log harvests, a quota on lumber will increase the price of lumber and reduce log harvests (see Fig. 3.3). This would then reduce the supply of jointly-produced products:

$$\frac{\partial S_i^*\left(P_i;P_{i-},P_j,P_{lum}\right)}{\partial P_{lum}} < 0, \quad \forall i = 1, \ldots, k. \tag{3.1}$$

Conversely, in markets for products that are competitive in production with lumber, an increase in the price of lumber will reduce the logs and sawmill residues available for those products and thus reduce their supply. It will also increase the substitution of wood and other products for lumber. For example, T-joists that employ OSB and less lumber will substitute for 2 in × 12 in joists – the supply of such substitutes will increase. Overall, it is unclear how changes in the price of lumber will impact the production of joint products that are also substitutes in consumption:

$$\frac{\partial S_j^*\left(P_j;P_{j-},P_i,P_{lum}\right)}{\partial P_{lum}} \lessgtr 0, \quad \forall j = k+1, \ \ldots, K. \tag{3.2}$$

However, because lumber remains the most valuable use of logs, it is unlikely that the higher price of lumber will increase the supply of wood products that substitute for lumber in consumption.

One can measure the welfare implications of policy intervention in multiple vertically and horizontally connected markets. Again we assume that the prices of inputs into the production of logs do not change with changes in the harvests of logs. Likewise, we assume that changes in the supplies of the *K* outputs produced from wood fiber, whether lumber, plywood or wood pellets, do not change the prices in the downstream markets in the vertical chain – the demand functions in these markets are perfectly elastic. For example, as the global supply of wood pellets changes, the prices received for electricity in various countries are not impacted. Likewise, as the supplies of lumber and/or plywood change, the prices of residential construction or furniture do not change. Therefore, we ignore these upstream and downstream markets. The supply curves for the two markets behave in a manner consistent with equations (3.1) and (3.2).

By assumption, the market for logs is the common and only upstream market – logs are the main input into the processing of forest products (although it is unlikely that trees are harvested in the absence of lumber demand). In Fig. 3.5, there exist two markets horizontal to lumber in the supply chain. The first is market *j* in panel (a), whose output is considered competitive for fiber and a substitute for lumber in construction, say (e.g. T-joists employing OSB substitute for 2 in × 12 in beams). The higher price of lumber will shift the supply curve for substitutes in consumption outwards, but the fact that fewer logs are harvested tends to shift the supply curve inwards. In Fig. 3.5(a), we show the supply curve as shifting inwards, because lumber remains the most valuable use for logs so that OSB and other products take second stage to lumber.

Second, market *i* represents commodities that are considered a joint product with lumber; outputs *i* utilize fiber that originates with the manufacture of lumber and/or constitute products that complement lumber in consumption (e.g. density fiberboard in doors). When lumber production falls, the supply function for products *i* shifts to the left, both because commodities *i* are jointly produced with lumber and because they are complements with lumber in final consumption – the less lumber produced, the higher is its price and the less of commodity *i* is supplied.

How do we measure welfare across various horizontal markets? In our discussion of a vertical chain, we found that the welfare costs of a lumber quota

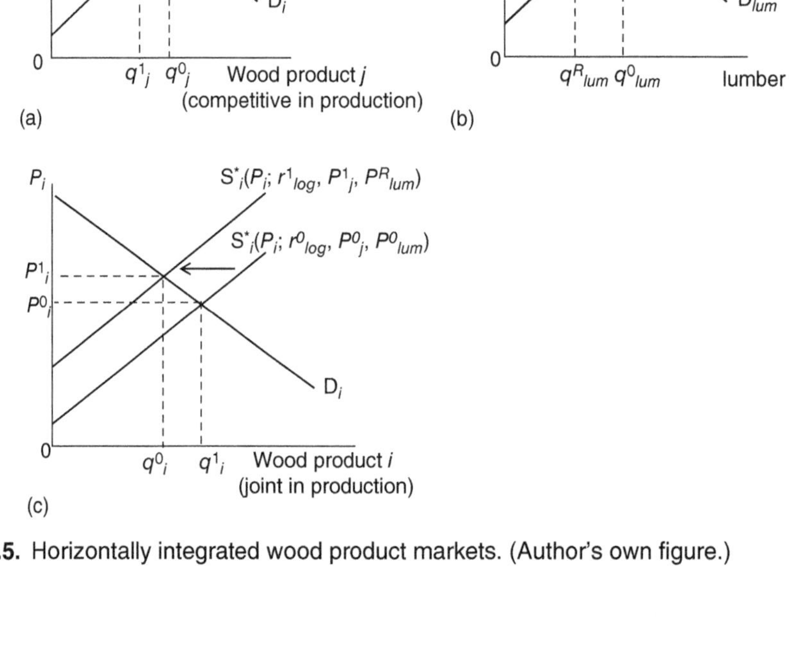

Fig. 3.5. Horizontally integrated wood product markets. (Author's own figure.)

could be measured in the upstream log and downstream product markets. We could measure the consumer surplus losses as the sum of the producer surpluses across all of the immediately downstream markets. Of course, it would be easier to deal with only the market impacted by the policy and measure the consumer welfare changes, actually the quasi-rents accruing in the downstream markets, as the area under the demand function for lumber, namely, area (*a*+*b*). Likewise, the quasi-rent lost in the lumber market, area (*c*+*d*), can be measured in the upstream log market as discussed with respect to Fig. 3.3. The only welfare area that is not considered is the lost quasi-rent in the log market, which is given by area (*u*+*v*) in Fig. 3.4, although area *u* constitutes an income transfer from loggers (or, more appropriately, forestland owners) to consumers of logs – it is part of the quasi-rent accruing to lumber producers and thus measured in that market. Thus, only the deadweight loss given by area *v* needs to be considered; as discussed above, it can be measured using the changes in the prices and outputs of logs, without the need to model explicitly the log market supply and demand functions. Indeed, if one suspects that area *v* is relatively small compared with the other welfare measures, it might be possible to ignore the log market and thus measure everything in the lumber market.

In the analysis of horizontal markets, we have so far examined wood products only – products that rely on wood fiber directly from logs or indirectly from sawmill residuals and logging residues (if it pays to gather and haul logging residues to market). The products under consideration were complementary to lumber in final demand markets and joint in production, or substitutes for lumber in consumption and competitive in production (competing for the same wood fiber). However, there are other products that are in the horizontal chain. For example, concrete, aluminum and steel are substitutes for lumber in construction, while roofing shingles are complements. An increase in the price of lumber will shift the demand for substitutes outwards, while it reduces the demand for complements. Do we need to construct supply and demand functions for these related markets in order to measure welfare changes in the horizontal chain?

There is nothing to measure in horizontally affected markets for complements or substitutes unless there is a distortion (market failure) in the market for a complementary or substitutable good, in which case welfare changes in such markets need to be taken into account. A distortion occurs in a market when price (P) exceeds the marginal cost (MC) of production. Price represents the marginal value of the commodity. Whenever P>MC, society places a higher value on the item than what it costs to produce it. Therefore, society should continue producing more of the good or service in question until its marginal value equals its cost of production. The demand for concrete increases when a quota is imposed in the lumber market as the price of lumber has risen. If the concrete market is perfectly competitive so that P=MC before and after the increase in demand caused by the higher price of lumber, there is no welfare loss or gain to measure in the market for concrete. If P>MC in the concrete market, the increase in demand for concrete leads to an increase in welfare that is set against any loss in welfare in the lumber market. However, the increase in welfare in the concrete market could never offset the loss in welfare in the lumber market, because, in that case, the initial allocation of resources (how people spent their money) would not have been efficient to begin with, contrary to the assumptions of welfare measurement.

In dealing with such horizontal markets, it is best to adopt the following rule for measuring welfare. Suppose that a government policy affects market 1. Then, to the welfare gain (loss) measured in market 1, one must add (subtract) the indirect surpluses that are measurable in related but distorted markets. The total welfare effect, ΔW, can be decomposed into its direct plus indirect welfare areas. Assuming fixed producer prices, the change in welfare can be determined as follows (Boadway and Bruce, 1984, pp. 252–255):

$$\Delta W \approx \int_{q_1^0}^{q_1} \left[D(q) - S(q)\right] dq \pm \sum_{j=2}^{n} \left[P_j - MC_j\right] \Delta q_j, \tag{3.3}$$

where q_1^0 is the initial consumption in market 1 before the policy and q_1 is the final consumption, and Δq_j is the change in the jth market that is a substitute or complement of good 1. The first term in the above formula is the direct effect and constitutes the total willingness to pay (area under demand function) minus the total cost of production (area under supply function) in going from q_1^0 to q_1. The above measure of welfare change is only approximate, however, because a correct measure would require a great deal more information that is rarely available to the policy analyst.

3.2 Background: Trade Modeling

Despite their usefulness for evaluating policy, analytic models have deficiencies that can only be addressed with an appropriate numerical model. Because a country's domestic forest product sector is inevitably linked to international markets, economic policies related to forestry policies must be viewed in the context of their impacts on foreign markets. Not only is the forest products industry connected through international trade, it also comprises many interconnected wood products. As wood fiber is generally sourced from the initial harvest of logs, the manufacturing of secondary wood products will be affected not only by the supply of logs, but also by competition for residual fiber. Indeed, the initial demand for logs is derived from the demand functions for various manufactured wood products, including lumber. Any structural shifts in the market for one of these products will inevitably impact the others. Numerical modeling is required if one wants to consider the impact of forestry policies on prices, quantities and wellbeing across various regions.

Economists at the Canadian Forest Service's Pacific Forestry Centre in Victoria have worked alongside academic economists to develop a spatial price equilibrium (SPE) modeling platform for analyzing forest sector trade between the U.S. and Canada. The modeling platform actually consists of various models that have looked at different aspects of global forest trade. Some models have focused solely on various aspects of U.S.–Canada softwood lumber trade

(Delcourt, 1995; Mogus *et al.*, 2006; Abbott *et al.*, 2009), while others have taken a more nuanced global perspective that included commodities other than just softwood lumber – a vertical chain with a horizontal chain (van Kooten and Johnston, 2014; Johnston and van Kooten, 2016, 2017). Because these modeling efforts employ a similar approach and many of the same researchers, we refer to this collection of models as the REPA trade model, referring to the Resource Economics and Policy Analysis research group located in the Department of Economics at the University of Victoria (see http://web.uvic.ca/~repa/, accessed September 4, 2020). Other researchers have used a similar SPE approach to model forest products trade (Holland and Judge, 1963; Takayama and Judge, 1971; Haynes *et al.*, 1978; Sedjo and Lyon, 1983; Perez-Garcia, 1993; Toppinen and Kuuluvainen, 2010; CINTRAFOR, 2014).

In the SPE framework, the modeler needs to have data on production, consumption, prices, and domestic supply and demand elasticities to construct the demand and supply functions in each region, plus shipping and handling (S&H) costs and bilateral trade among regions. The objective in SPE models is to maximize the overall welfare across all regions, as represented by the sum of all consumer and producer surpluses plus any resource- and policy-induced rents. Linear supply and demand functions are generally assumed for each region/country. With linearity, welfare measures are quadratic. Thus, the SPE model constitutes a quadratic program that is solved in a standard way – maximizing the quadratic objective function subject to various biophysical and economic constraints. The advantage of linear and quadratic programming models compared with nonlinear programming models is that they can be solved using the simplex algorithm (Paris, 1991; Elwes, 2012). A lumber quota (or even a per unit export tax or import tariff) can be implemented quite simply by adding a restriction on production of lumber from regions affected by the quota (or adding a per unit cost that represents the tax). However, if an ad valorem tax/tariff is employed, it is necessary to use mixed complementarity programming (MCP) due to endogeneity between price and quantity (see Howitt, 1995; Nicholson and Bishop, 2004).

The most recent version of the REPA trade model accommodates the 2006 SLA's two-tiered tariff regime (Johnston and van Kooten, 2017). A lower ad valorem tax or tariff was applied to sales below each Canadian region's quota limit (except for Atlantic Canada), after which the ad valorem tax is increased to a rate that was designed to prevent further sales. Because we investigate the impact of an ad valorem tax on lumber, the tax revenue and quota rent are dependent on the effect that the tax has on prices and quantities. Since price and quantity are simultaneously determined in the model, a standard linear programming approach for solving the SPE model cannot be employed as would be the case with a hard quota. Rather, as noted above, it is necessary to employ MCP, and then within the context of a tariff rate quota regime (Nicholson and Bishop, 2004). A mathematical representation of the subsequent MCP model with TRQ is found in Chapter 4 (this volume) and in Johnston and van Kooten (2017).

Here we provide a brief description of the underlying fiber flows that form the basis of the extended multiple-product, 20-region trade model. Harvests of timber lead to a supply of industrial roundwood that provides the fiber for a number of downstream products: sawnwood (lumber), plywood, particleboard (waferboard and OSB), fiberboard (medium density and hardboard), wood pulp, and wood pellets. Although lumber and plywood are the most lucrative uses of roundwood, their production also provides residuals in the form of chips and sawdust that can be used to produce fiberboard, pulp and wood pellets as indicated in Fig. 3.6. Finally, the harvest and processing of industrial roundwood from the initial harvest produces residuals (roadside debris, tree tops, branches, etc.), which may also be used in the production of wood pellets (although this is not done here, because transportation costs are often too great).

The model assumes that, while changes in countries' forest policies will affect prices of forest products, they have no discernible impact on the relative prices of goods and services elsewhere in the economy. Since it is a spatial price equilibrium trade model, it is assumed that, in the absence of trade barriers and transaction costs, prices would be the same in every region as a result of spatial arbitrage – the law of one price (LOP) holds (Chapter 2 in Vercammen, 2011). Differences in prices between regions are thus assumed to be the result of transaction costs, and include costs associated with S&H (e.g. freight, insurance, exchange rate conversion fees), plus tariffs and other non-tariff barriers.

In the model, Canada is divided into five regions – Atlantic Canada, Central Canada, Alberta, BC Interior, and BC Coast. The United States is divided into three regions (South, North, West), and Asia is separated into China, Japan, and Rest of Asia. Chile, Australia, New Zealand, Finland, Sweden, and Russia are also separate regions, while the remaining regions comprise Rest of Europe, Rest of Latin America, and the Rest of the World (ROW). The model calculates production of logs and various wood products and their consumption in each region, and associated bilateral regional trade flows. It is solved numerically in an integrated Excel-R-GAMS environment.

The REPA forest trade model is described in greater detail in Chapter 4, while examples of its use are provided in Chapter 5, this volume. In both chapters, various General Algebraic Modeling System (GAMS) codes are provided. An alternative model to the REPA platform is the Global Forest Products Model (GFPM). It relies on more general trade relations – each country trades with the

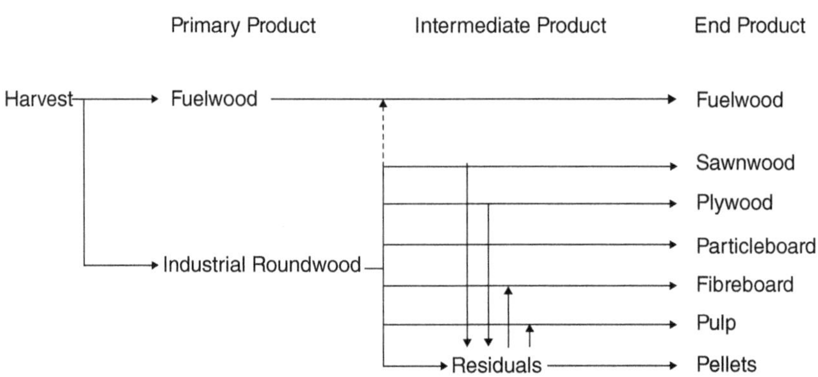

Fig. 3.6. Forest Product Flow Chart. (Source: Johnston and van Kooten, 2014.)

Rest of the World, but not with other countries (Buongiorno *et al.*, 2003; Sun *et al.*, 2010). That is, GFPM sacrifices information on bilateral flows for greater product detail. The GFPM model is described in Chapter 6, this volume.

References

Abbott, B., Stennes, B. and van Kooten, G.C. (2009) Mountain pine beetle, global markets and the British Columbia forest economy. *Canadian Journal of Forest Research* 39(7), 1313–1321.

Boadway, R.W. and Bruce, N. (1984) *Welfare Economics*. Blackwell Publishing, Oxford, UK.

Buongiorno, J., Zhu, S., Zhang, D., Turner, J. and Tomberlin, D. (2003) *The Global Forest Products Model: Structure, Estimation and Applications*. Academic Press/Elsevier, San Diego, California.

CINTRAFOR (2014) Center for International Trade in Forest Products. University of Washington, School of Environmental and Forest Sciences, Seattle, WA. Available at: https://www.cintrafor.org/data.html (accessed July 20, 2020).

Delcourt, G.V. (1995) Global softwood lumber trade: a spatial equilibrium model. MSc thesis. University of British Columbia, Vancouver, Canada.

Elwes, R. (2012) The world maker: is time running out for the clever piece of maths that runs modern life? *New Scientist* 11, 33–37.

Haynes, R.W., Holley, D.L. and King, R.A. (1978) A recursive spatial equilibrium model of the softwood timber sector. *Technical Report No. 57*. School of Forest Resources, North Carolina State University.

Holland, I.I. and Judge, G.G. (1963) Estimated interregional flows of hardwood and softwood lumber. *Journal of Forestry* 61, 488–492.

Howitt, R.E. (1995) Positive mathematical programming. *American Journal of Agricultural Economics* 77, 329–342.

Johnton, C.M.Y. and van Kooten, G.C. (2014) Economic consequences of increased bioenergy demand. *The Forestry Chronicle* 90(5), 636–642.

Johnston, C.M.T. and van Kooten, G.C. (2016) Global trade impacts of increasing Europe's bioenergy demand. *Journal of Forest Economics* 23, 27–44.

Johnston, C.M.T. and van Kooten, G.C. (2017) Impact of inefficient quota allocation under the Canada–U.S. softwood lumber dispute: a calibrated mixed complementarity approach. *Forest Policy and Economics* 74, 71–80.

Just, R.E., Hueth, D.L. and Schmitz, A. (2004) *The Welfare Economics of Public Policy. A Practical Approach to Project and Policy Evaluation*. Edward Elgar, Cheltenham, UK.

Mogus, A., Stennes, B. and van Kooten, G.C. (2006) Canada–U.S. softwood lumber trade revisited: examining the role of substitution bias in the context of a spatial price equilibrium framework. *Forest Science* 52(4), 411–421.

Nicholson, C.F. and Bishop, P.M. (2004) *U.S. dairy product trade: modeling approaches and the impact of new product formulations*. Final Report for NRI Grant #2001-35400-10249, March 2004.

Paris, Q. (1991) *An Economic Interpretation of Linear Programming*. Iowa State University Press, Ames, Iowa.

Perez-Garcia, J.M. (1993) Global forestry impacts of reducing softwood supplies from North America. *CINTRAFOR Working Paper 43*. University of Washington, School of Environmental and Forest Sciences, Seattle, Washington.

Sedjo, R. and Lyon, K. (1983) Long-term forest resources trade, global timber supply, and intertemporal comparative advantage. *American Journal of Agricultural Economics* 65(5), 1010–1016.

Sun, L., Bogdanski, B., Stennes, B. and van Kooten, G.C. (2010) Impacts of tariff and non-tariff trade barriers on global forest products trade: an application of the Global Forest Products Model. *International Forestry Review* 12, 49–65.

Takayama, T. and Judge, G. (1971) *Spatial and Temporal Price and Allocation Models*. North-Holland, Amsterdam, Netherlands.

Toppinen, A. and Kuuluvainen, J. (2010) Forest sector modelling in Europe – the state of the art and future research directions. *Forest Policy and Economics* 12, 2–8.

van Kooten, G.C. and Johnston, C.M.T. (2014) Global impacts of Russian log export restrictions and the Canada–U.S. lumber dispute: modelling trade in logs and lumber. *Forest Policy and Economics* 39, 54–66.

Vercammen, J. (2011) *Agricultural Marketing. Structural Models for Price Analysis*. Routledge, London and New York.

4 Modeling Bilateral Forest Products Trade

Craig M.T. Johnston[1], Brad Stennes[2] and G. Cornelis van Kooten[3]

[1]Bank of Canada, Ottawa, ON Canada; [2]Canadian Forest Service's Pacific Forestry Centre, Victoria, BC Canada; [3]University of Victoria, Victoria, BC Canada

Abstract

The focus in this chapter is on the development of mathematical programming models used to model bilateral forest products trade. Theoretical outlines are provided of a multi-region, single product trade model and of an integrated, multi-region, multi-product trade model. The objective function and constraints are described mathematically, while the analysis takes into account horizontal and vertical chains and the need to calibrate the model using observed trade flows. Data sources are discussed, and the GAMS code is provided for the uncalibrated and calibrated versions of the model.

The Canada–U.S. softwood lumber dispute is the *raison d'être* for much applied work in modeling forest products trade, especially on Canada's side. In this chapter, we examine several spatial price equilibrium (SPE) trade models that are currently used to investigate the implications of trade barriers imposed on Canadian exports of softwood lumber to the United States. The reason we consider bilateral trade is so that we can determine the impacts of trade restrictions on various regions in North America. We begin in the next section by specifying a general but vertically integrated SPE trade model.

4.1 Integrated Vertical and Horizontal Chain Model Specification

In this section, we develop a general spatial price equilibrium trade model that assumes forestry policy aims to prevent exports of logs from one or more jurisdictions. The SPE model tracks the use of logs in the production of lumber and other wood products (see Fig. 3.6 in Chapter 3, this volume), and aims to determine the price, quantity, and welfare policy implications across regions and products. The policy instrument is assumed to be a simple per unit tax on exports.

Doi: 10.1079/9781789248234.0004

4.1.1 Objective function

Consider first the wood processing sector. Each region is assumed to have a set of linear (inverse) demand and supply curves for each downstream product k:

$$P_d^k = \alpha_d^k - \beta_d^k q_d^k, \alpha_d^k, \beta_d^k \geq 0, \forall d = 1,\ldots,M, \forall k, \quad (4.1)$$

and

$$P_s^k = a_s^k + b_s^k q_s^k, a_s^k, b_s^k \geq 0, \ \forall s = 1,\ldots, N, \forall k, \quad (4.2)$$

where $k \in$ {lumber, plywood, particleboard, fiberboard, pulp, wood pellets}, q_d^k refers to the quantity of commodity k consumed in demand region d, and q_s^k refers to the quantity of wood product k produced by supply region s. For convenience, we use d to denote a net demand region and s a net supply region, although a region may be simultaneously a supplier and demander of the commodity in question. There are M demand (import) regions and N supply (export) regions.

One objective of the forest trade model is to maximize the sum of the consumer and producer surpluses across all relevant wood-processing sectors. The consumer and producer surpluses are found by maximizing the sum of the areas under the M demand schedules (4.1) and subtracting the sum of the areas under the N supply schedules (4.2). These respective areas are given by:

$$B_d^k = \int_0^{q_d^k} \left(\alpha_d^k - \beta_d^k x\right) dx = \alpha_d^k q_d^k - \tfrac{1}{2}\beta_d^k q_d^{k^2}, \text{ and} \quad (4.3)$$

$$C_s^k = \int_0^{q_d^k} \left(a_s^k + b_s^k x\right) dx = a_s^k q_s^k + \tfrac{1}{2} b_s^k q_s^{k^2}, \quad (4.4)$$

where x is an integration variable, B_d^k is the total benefit (area under the demand function) in demand region d for product k, and C_s^k is the total cost (area under the marginal cost/supply function) in supply region s for product k (see Vercammen, 2011, p. 22). Given lack of good data, a supply elasticity of one is often assumed, which implies that the supply schedules pass through the origin. Vercammen also provides the welfare equation if supply schedules intersect the horizontal axis so that they have negative intercepts.

Now consider the markets for industrial roundwood (pulp logs and coniferous logs). As noted earlier, the demand for logs is a derived demand that depends on the production of downstream lumber, plywood, pellets, pulp, etc. For each wood product k, its derived demand is given by its output price multiplied by the marginal physical product of the input (logs) in the production of the kth commodity: $P^k \times MP_{logs \to k}$. The total derived demand for logs is given by the horizontal sum of the individual k derived demands for logs. However, as shown in Chapter 3, the change in consumer surplus in the log market caused by a policy shock affecting logs can be evaluated in the downstream markets as the sum of the changes in the producer surpluses (quasi-rents) in the downstream wood processing markets – changes in the consumer surplus in the log market are measured by the changes in producer surpluses in the downstream markets. Thus, it is necessary to include in the objective function only the producer surplus in the log market. Assume that the supply (marginal cost) of logs in region u is linear: $r_u = m_u + n_u Q_u$, m_u, $n_u \geq 0$, where Q_u is the quantity of logs in country u. The producer surplus from supplying logs from any region u is:

$$QR_u = r_u Q_u - \int_0^{Q_u} (m_u + n_u x)dx = \tfrac{1}{2} n_u Q_u^2, \forall u = 1,\ldots,U, \tag{4.5}$$

where U regions supply logs.

The overall objective in the forest trade model is to maximize the sum of the necessary producer and consumer surpluses provided above, while subtracting the shipping and handling costs and associated taxes. The objective function to be maximized can thus be written as:

$$\begin{aligned} W = \textstyle\sum_{k=1}^{K} \left[\sum_{d=1}^{M} B_d^k - \sum_{s=1}^{N} C_s^k - \sum_{d=1}^{M} \sum_{s=1}^{N} \left(t_{s,d}^k + \tau_{s,d}^k \right) q_{s,d}^k \right], \\ + \textstyle\sum_{u=1}^{U} \left[QR_u - \sum_{s=1}^{N} \left(\delta t_{u,s} + \tau_{u,s} \right) Q_{u,s} \right] \end{aligned} \tag{4.6}$$

where W refers to the overall global wellbeing from trade in forest products, $\tau^k_{s,d}$ is the cost (\$/m^3) of transporting processed forest product k from supply region s to demand region d, and $\delta t_{u,s}$ is the cost of transporting industrial roundwood (logs) from region u to region s, where δ is a parameter that takes into account the extra cost of transporting logs because they occupy more space per cubic meter than lumber (whose cost of transport from region u to region s is given by $t_{u,s}$). Finally, $\tau_{u,s}$ is the tax on logs (\$/m^3) originating in log supply region u and sold to wood product producing region s, while $\tau^k_{s,d}$ is the tax on wood product k originating in supply region s and exported to region d.

4.1.2 Constraints

Objective (4.6) is maximized subject to a series of biophysical and economic constraints relating to the availability of timber harvests, log supply and wood product manufacturing limits. The essential constraints are material flows and productivity constraints that ensure total supply equals total demand for each region/country and each product. The model constraints are summarized as follows. First, the quantity of industrial roundwood of each type $L \in$ {saw logs, veneer logs, pulpwood logs} = {SL, VL, PL} produced by any log producing region u must be no greater than its harvest of logs (h_u), and the region's ability to convert harvested timber into various industrial roundwood components:

$$Q_u = \sum_{L \in \{SL,VL,PL\}} Q_u^L \leq \varphi_u^L \times h_u, \forall u. \tag{4.7}$$

The parameter φ_u^L indicates how much coniferous industrial roundwood of each type is recovered from the timber harvest in region u, which depends on size and species of trees, as well as the region's technical skills, capital, and other factors. The aggregate of the various log types in region u is denoted Q_u.

The sale of logs by region u to log consuming regions s, including domestic sales, must not exceed the total supply of logs in region u:

$$\sum_{s=1}^{N} Q_{u,s}^{L} \leq Q_{u}^{L}, \forall\ u,\ L. \tag{4.8}$$

The quantity of logs supplied to region s must be greater than or equal to the amount required for the production of downstream wood products:

$$\sum_{u=1}^{U} Q_{u,s}^{L} \geq Q_{s}^{L}, \forall\ s,\ L. \tag{4.9}$$

Logs are used as inputs into the production of the K downstream wood products. It follows that the sale of downstream wood products from supplying region s to all consuming regions must be no larger than what is produced in region s:

$$\sum_{d=1}^{M} q_{s,d}^{k} \leq q_{s}^{k}, \forall\ s,\ k. \tag{4.10}$$

Similarly, the supply of downstream products from all supply regions to region d, and including domestic supply, must be greater than or equal to the demand of region d:

$$\sum_{s=1}^{N} q_{s,d}^{k} \geq q_{d}^{k}, \forall\ d, k. \tag{4.11}$$

We do distinguish between primary and secondary products processed from logs in our model - our defined primary products are generally more valuable than our secondary products (but not always). Secondary product may be more valuable depending on the relative quality of logs and location of plant facilities. Consider, as an example, fast-growing pine plantations located next to a power plant; the pine is grown primarily to be used as a biomass fuel. Alternatively, such trees might be best used to produce pulp if no sawmill is in the vicinity.

Lumber and plywood must necessarily be considered primary products for sawlogs (*SL*) and veneer logs (*VL*), while wood pulp is the primary product from pulp logs (*PL*). In addition, secondary products (particleboard, fiberboard, and wood pellets) can employ wood fiber from logs in direct competition with the primary products, with wood pulp also considered a secondary product when it comes to non-pulp logs. Therefore, we denote $f \in$ {particleboard, fiberboard, pulp, pellets}$\subset K$ and $nf \in$ {lumber, plywood} $\subset K \ni f \cup nf = K$.

Secondary products rely on chips and residuals from sawmilling and plywood manufacture for the most part. For simplicity, however, we assume that industrial roundwood gets allocated to each of our six (primary plus secondary) products so that all of the roundwood is utilized. This can be described using the following relation:

$$q_{s}^{k,L} \leq \rho_{s}^{k,L} \times \eta_{s}^{k,L} \times Q_{s}^{L}, \forall k, s, L. \tag{4.12}$$

In equation (4.12), the total available output in processing region s of wood product k from logs of type L, $q_{s}^{k,L}$, is determined by the proportion of the logs of type L used to produce k, denoted $\rho_s^{k,L}$, multiplied by the recovery factor, $\eta_s^{k,L}$, that converts logs into product and the amount of logs of type L available in region s. To ensure that all of the wood fiber is fully used we require

$$\sum_{k=1}^{K} \rho_s^{k,L} = 1. \tag{4.13}$$

The manufacture of lumber and plywood results in chips and other residuals (sawdust, planer shavings, residues) that are joint products that can be used to produce particleboard, fiberboard, wood pulp, and wood pellets. The total amount of wood chips and residuals produced in region s depends on the production of lumber and plywood, and can be determined from the following relation:

$$R_s^z = \sum_{nf}\left(q_s^{nf} \times v_s^{nf,z}\right), \forall\, s,\, z \in \{\text{wood chips,other residuals}\}, \tag{4.14}$$

where R^z_s is the amount of z (wood chips, sawdust, planer shavings or other residuals) produced in region s and $v_s^{nf,z}$ is the region's ability to recover residual z from each of the sawmilling and plywood manufacturing sectors.

The production of products f directly from logs in region s is denoted $q_s^{f,L}$ and is determined from equation (4.12). In addition, we find the amount of product f produced from chips and residual fiber using the following relationship:

$$q_s^{f_R} = \Sigma_z (w_{s,f}^z \times \theta_f^z \times R_s^z), \forall s, \tag{4.15),}$$

where $q_s^{f_R}$ denotes the quantity of wood product f produced from residual fiber (and not directly from logs). In addition, $w^z_{s,f}$ refers to the proportion of residual fiber of type z in region s that is used to produce product f, while θ_f^z is a parameter that converts residual fiber of type z into product f. The condition requiring that all residual fiber is exhausted is given by:

$$\Sigma_f\, w_{s,f}^z = 1, \forall s, z. \tag{4.16}$$

Pulp mills use chips from sawmilling and manufacture of plywood to the extent that such chips are not used for other products. OSB, particleboard, fiberboard and pellets can employ wood chips and other residuals from sawmilling and plywood production.

Finally, the total amount of product k produced by region s can now be determined as:

$$q_s^k = q_s^{f_R} + \sum_{L \in \{SL,VL,PL\}} q_s^{k,L}, \forall s, k, f \in k. \tag{4.17}$$

The constrained optimization program maximizes objective (4.6) subject to constraints (4.7) through (4.17), plus non-negativity conditions on the decision variables. For each of the relevant regions, the decision variables are the supply of industrial roundwood (sawlogs and veneer logs) and pulpwood (Q^L_u); bilateral flows of logs from supplying to wood processing regions ($Q^L_{u,s}$); production and consumption of product k in each region (q_s^k and q_d^k, respectively); and the bilateral trade flows of product k ($q^k_{s,d}$). The proportions of the logs of type L used to produce k ($\rho_s^{k,L}$), and the proportions of residual fiber of type z used in f, namely $w^z_{s,f}$, can also be determined endogenously in the model, although we have tended to let these be exogenously provided.

4.2 Calibrating the Integrated Vertical and Horizontal Chain Model

It is important that the forest trade model is calibrated so that the user can be confident that model projections are realistic. The calibration must be based on observed values and must be rooted in economic theory. Although trade models rely on observed data, it is often the case that computational deficiencies require an aggregation of firm and market characteristics. As a result, mathematical programming models of trade often experience extreme specialization in supply responses. As well, discrepancies between modeled and observed optimal values may arise due to mis-specified parameters, often originating from transaction costs per unit of product traded between two countries (e.g. non-tariff trade barriers). To deal with such problems, several calibration techniques have evolved.

One method is referred to as the historical mixes approach, which attempts to address the problem of extreme solutions (McCarl, 1982; Önal and McCarl, 1991). This approach is based on the fact that optimal solutions are often found at corners (extreme points), particularly when working with aggregate representative producers. (The simplex method that is used in solving linear and quadratic programming problems finds only corner solutions (Paris, 1991)). Since aggregation may bias the data, and hence the original problem, a region may be assigned a subset of possible production 'mixes' based on observed levels. This last point is justified as observed mixes must be optimal, or why would they have occurred in the first place? This calibration method takes historical choices (mixes) into account by constraining the current optimal values to be a weighted average of those observed choices. The weights may be determined endogenously within the mathematical programming framework, with the sum of the weights equaling 1. Chen and Önal (2012) extended this method by including decisions that are not historically observable. Simulated mixes of the 'new' decision variables are added to the historical mixes, allowing the optimization procedure to choose the weights, and again constraining the sum of the historical and synthetic weights to equal 1.

A second calibration method based on an approach originally proposed by Howitt (1995), referred to as positive mathematical programming (PMP), is increasingly applied to problems in agriculture and forestry (see de Frahan *et al.*, 2007; Paris, 2011, pp. 340–411; Heckelei *et al.*, 2012). Positive mathematical programming uses the notion that any calibration constraint can be represented in the objective function (e.g. a linear calibration constraint might be represented as a nonlinear cost function in the objective). Rather than adding arbitrary calibration constraints to ensure that the optimal solution to a mathematical program replicates what is observed (as in the historical mixes approach), the PMP method uses the shadow prices associated with such constraints to re-specify the objective function. The calibrated model is then solved to replicate the observed values exactly.

In trade models, elasticities of supply and demand along with observed production and consumption levels enable the construction of linear supply and demand functions. The greatest uncertainty relates to the transaction (including S&H) costs of moving goods from one region to another. This does not include the export taxes or import tariffs as these are usually known, considered separate from the transaction costs, and, in any event, constitute policy variables. Rather, it is the actual or realizable transaction costs that are unknown, but which can be determined using PMP as a calibration method. Thus, the PMP-calibrated transaction costs represent the 'effective' transaction costs between export and import regions. They are derived from the shadow prices on the calibration constraints relating to the observed bilateral flows of forest products (logs, lumber, wood pellets, etc.). Again the calibration is motivated by the fact that there is a discrepancy between the true transaction costs and the observed transaction costs, as determined by shipping, loading and unloading, insurance, and administrative costs, plus unaccounted-for tariffs and non-tariff barriers, such as phytosanitary standards. The main reason for this discrepancy occurs because (observed) transaction costs are measured with a significant degree of uncertainty (Paris *et al.*, 2011). To deal with and measure the hidden or unknown transaction costs (bribes, non-tariff barriers, etc.), one can utilize a two-stage positive mathematical programming model (Paris *et al.*, 2011).

The PMP specification in stage I maximizes objective (4.6) subject to constraints (4.7) through (4.17), with the addition of the following constraints:

Dual Variable

$$Q^L_{u,s} = \bar{Q}^L_{u,s} \qquad \left[\lambda^L_{u,s}\right] \tag{4.18}$$

$$q^k_{s,d} = \bar{q}^k_{s,d} \qquad \left[\lambda^k_{s,d}\right] \tag{4.19}$$

In this specification, it is assumed that we observe trade flows for industrial roundwood and k downstream wood products, $\bar{Q}^L_{u,s}$ and $\bar{q}^k_{s,d}$, respectively, along with their respective transaction costs $\delta t_{u,s}$ and taxes or tariffs $\tau^k_{s,d}$.

Upon obtaining the shadow prices $\lambda^L_{u,s}$ and $\lambda^k_{s,d}$ associated with the primal model, the objective function for stage II can be re-specified as follows:

$$\text{Maximize } W = \sum_{k=1}^{K}\left[\sum_{d=1}^{D} B^k_d - \sum_{s=1}^{S} C^k_s - \sum_{d=1}^{D}\sum_{s=1}^{S} T^k_{s,d} q^k_{s,d}\right] + \sum_{L\in\{SL,VL,PL\}}\sum_{u=1}^{U}\left[QR_u - \sum_{s=1}^{S}\delta T^L_{u,s} Q^L_{u,s}\right] \tag{4.20}$$

where $T^k_{s,d}$ now equals $t^k_{s,d} + \tau^k_{s,d} + \lambda^k_{s,d}$, and $T_{u,s}$ equals $t_{u,s} + \tau_{u,s} + \lambda^L_{u,s}$. In the second stage, the modified objective function (4.20) is maximized subject to the original constraints (4.7) through (4.17). With this modification, the model precisely duplicates the inter-regional fiber trade flows.

The fact that the shadow prices $\lambda^{L}_{u,s}$ and $\lambda^{k}_{s,d}$ can be negative indicates that the original transaction cost data fail to include missing information, perhaps even policy instruments such as export subsidies. Indeed, Paris *et al.* (2011) indicated that, in some instances, the overall effective transaction costs between two countries might even be negative, as when export subsidies are larger than the sum of other transaction costs. In some circumstances, this may provide additional insight into the potential restrictiveness of trade measures that are otherwise difficult to quantify, such as non-tariff trade barriers.

4.3 Economic Surplus and Income Redistribution

In Chapter 3, we identified the appropriate welfare measures used in trade models, where trade models consist of various products that are vertical and horizontal to the market impacted by a forestry policy. A forest policy might be an export tax on logs, a quota on lumber imports, or some other policy scenario. In addition to consumer surplus and quasi-rent (producer surplus), one must obtain estimates of resource rents, policy-induced quota rents and/or tax/tariff revenues. (In practice, some part of any quota rent might simply be squandered away through rent-seeking activities or simply wasted due to other inefficiencies, although we make no attempt to measure such waste.) In bilateral SPE trade models, the welfare measures and income transfers are usually calculated after the model has been solved for the optimal bilateral trade flows. The following equations provide the mathematical derivation of these welfare measures.

4.3.1 Wood processing sector

Consider first the *k* downstream wood processing markets in the vertical supply chain. The consumer surpluses in each of these markets and each commodity are given by:

$$CS^k_d = \int_0^{q^k_d} \left(\alpha^k_d - \beta^k_d x\right)dx - P^k_d q^k_d = \left(\alpha^k_d q^k_d - \tfrac{1}{2}\beta^k_d (q^k_d)^2\right) - \left(\alpha^k_d - \beta^k_d q^k_d\right)q^k_d \tag{4.21}$$

$$= \tfrac{1}{2}\beta^k_d (q^k_d)^2, \forall s,k,$$

where P^k_d is the demand price for product *k* in the domestic market, and q^k_d is the quantity of product *k* consumed. Likewise, the producer surpluses or quasi-rents in these *k* downstream markets are given by:

$$QR^k_s = P^k_s q^k_s - \int_0^{q^k_s} \left(a^k_s + b^k_s x\right)dx = \left(a^k_s + b^k_s q^k_s\right)q^k_s - \left(a^k_s q^k_s + \tfrac{1}{2} b^k_s (q^k_s)^2\right) \tag{4.22}$$

$$= \tfrac{1}{2} b^k_s (q^k_s)^2, \forall s,k,$$

where P^k_s is the supply price for product *k* in the domestic market, and q^k_s is the quantity of product *k* that is produced.

It might be worth recalling that, in principle, a region might only be a supply or demand region, but in practice regions are usually both suppliers and demanders of each of the *k* products. Further, given that the regions in the model are quite large, each supplies some amount of harvested timber to its domestic market for processing.

In each of the downstream markets, a variety of distortions might exist. These consist of tariff and non-tariff trade barriers, export taxes, illegal fees, quotas, and so on. These distortions to trade are captured in the PMP calibration process so that they are included in the revised S&H and other transaction costs. However, when we examine the impact of various policies (e.g. tariff or export tax, quota), income transfers will occur and these can be measured in two ways. First, with tariffs or taxes, the income accrues to government and is calculated simply as the quantity affected (traded or sold) multiplied by the tariff/tax rate. Second, some policies create distortions that result in a wedge between the demand price and the marginal cost (supply price). This leads to a policy-induced scarcity rent that is calculated as follows:

$$\begin{aligned} SR_y^k &= \left(P_y^{k,Demand} - P_y^{k,Supply}\right)\bar{q}_y^k \\ &= \left\{\left(\alpha_y^k - \beta_y^k \bar{q}_y^k\right) - \left(a_y^k + b_y^k \bar{q}_y^k\right)\right\}\bar{q}_y^k, \forall k, y \in \{s,d\}, s \neq d, \end{aligned} \quad (4.23),$$

where $\bar{q}_y^k$ refers to the quantity of *k* consumed in market *y*. In essence, since the producer surplus calculated in equation (4.22) does not include the policy-induced scarcity rent, it is necessary to include this scarcity rent as a transfer, although it is not clear who captures it; it is simply determined by the size of the wedge between the demand and supply prices multiplied by the quantity sold in market *y* ($\bar{q}_y^k$).

4.3.2 Upstream Log Markets

Now turn to the market for logs. As noted earlier, because the demand for logs is a derived demand, the consumer surpluses in log markets are measured as quasi-rents in the *K* downstream markets (lumber, plywood, OSB, wood pellets, etc.). It is necessary, therefore, only to measure the quasi-rent in the log markets. The best measure of the quasi-rent in any log market is given by equation (4.5) and is similar to equation (4.22); it is given by:

$$QR_u^L = \frac{1}{2} n_u^L \left(Q_u^L\right)^2, \forall u, L, \quad (4.24)$$

where n_u^L is the slope of the type-*L* log supply curve in region *u*.

To this must be added any scarcity rent associated with resource scarcity or some policy that creates a wedge between the demand and supply price of logs, thereby creating a policy-induced rent. Because we do not explicitly include demand functions for logs in the trade model, we rely on the shadow price of logs. The shadow price of logs gives the addition to global wellbeing, as defined in objective function (4.6), if an additional log were available. Therefore, policy-induced rent plus the rent from resource scarcity in the log market can be calculated by the shadow price of logs times the volume produced:

$$SR_u^L = \lambda_u^L \bar{Q}_u^L, \forall u, L, \tag{4.25}$$

where λ_u^L is the shadow price and $\bar{Q}_u^L$ is the equilibrium production of logs of type L in region u.

The surpluses in equations (4.21) through (4.25) are summed to obtain the total surplus from trade.

4.4 Forest Trade Model Specifications

In the previous sections, we described a general SPE trade model that takes into account vertical and horizontal chains. Here we discuss several models that have been developed to examine various aspects of the softwood lumber trade dispute between Canada and the U.S., as well as policies implemented outside North America, specifically a European policy to increase generation of electricity from wood biomass. However, we also consider models that are more encompassing in terms of the regions explicitly modeled and the number of commercial forest products. Of course, the simplest model would focus only on softwood lumber trade between Canada and the United States, with a single 'rest of the world' region included to make the model tractable. The reason for this narrow focus is twofold. First, the main reason for harvesting timber is to produce lumber, with processing of other wood products dependent on activities related to sawmilling. Harvest levels would be much reduced if lumber was not the main objective; it is too costly to harvest logs only to produce wood products other than lumber. Second, the vast majority of Canada's softwood lumber exports are to the U.S. (as indicated in Chapter 2), although other countries may be modeled so as to evaluate the potential of markets outside the U.S. (e.g. China, Japan). Depending on its purpose, the rest of the world might be aggregated into a single region, or might be considered with greater disaggregated detail.

In Appendix 4.A, we provide the GAMS code for a straightforward model of softwood lumber trade that disaggregates Canada into five regions and the U.S. into three regions, and aggregates countries outside North America into 13 regions rather than one, thereby resulting in a 20-region bilateral SPE trade model. The data required to run the model and model outputs are also provided in the appendix. Then, in Appendix 4.B, we provide the GAMS code required to calibrate the basic model, but, unlike in Appendix 4.A, no model outputs are provided. Finally, a trade model that has the same 20 regions but now with eight forest products (see Fig. 3.6, this volume) is provided in

Appendix 4.C. The actual GAMS code for these three models as well as some additional GAMS code for other forest trade models, along with the data needed to run them, are provided under the 'Trade Models' tab at https://www.vkooten.net/ (accessed September 4, 2020).

The general mathematical programming formulation for these models is given by:

Maximize

$$W = \sum_{d=1}^{D}(\alpha_d - 0.5\beta_d q_d)q_d - \sum_{s=1}^{S}(a_s + 0.5b_s q_s)q_s - \sum_{s=1}^{S}\sum_{d=1}^{D}(t_{s,d} - \tau_{s,d})x_{s,d}$$

Subject to:

$$\sum_{d=1}^{D} x_{s,d} \le q_s \text{ (sales cannot exceed production)}$$

$$\sum_{s=1}^{S} x_{s,d} \ge q_d \text{ (production must exceed demand)}$$

$$x_{s,d} = x^0_{s,d} \text{ (calibration constraint) } [\lambda_{s,d}]$$

$$x_{s,d},\ q_s,\ q_d \ge 0 \text{ (non-negativity)}$$

In the above specification, $x_{s,d}$ refers to the trade between supply region s (of which there are S) and demand region d (of which there are D); q_s is the total production of region s; and q_d is the total demand in region d. As before, α_d and β_d are the parameters of the demand function in region d, while a_s and b_s are the marginal cost parameters for supply region s. Finally, $t_{s,d}$ and $\tau_{s,d}$ refer to the S&H and other transaction costs and taxes, respectively, that exist in softwood lumber trade between regions s and d.

The shadow prices associated with the calibration constraints, $\lambda_{s,d}$, can be positive or negative, as the constraint is an equality rather than an inequality. The shadow prices are then used to modify the transaction costs as follows: $t'_{s,d} = t_{s,d} + \lambda_{s,d}$, where $t'_{s,d}$ are the calibrated transaction costs to use in the second stage of the program (see Appendices 4.B and 4.C).

4.5 Model Data

Our discussion of model data relies on and updates Johnston and van Kooten (2016), who employed a multiple-product, 20-region model to examine an increase in European demand for wood pellets to produce electricity. Data availability is the greatest challenge that the modeler faces, with the degree of difficulty enhanced where country-level data need to be disaggregated into regions. Disaggregating national consumption data can be done using proportions of regional population, incomes or housing starts, while information on regional shares of national softwood lumber production is usually available from domestic statistical agencies. Aggregating national consumption and production data from the Food and Agriculture Organization (FAO) of the United Nations (UN) into regions is somewhat easier. (We rely on UN data from FAO and the UN's ComTrade Database, available at https://comtrade.un.org/, accessed September 4, 2020, aggregating country-level and bilateral trade data into pre-specified regions using the 'tidyverse' package in R.)

The greatest challenges are finding appropriate prices and elasticities of supply and demand, particularly at the sub-national (regional) level. This information is needed along with production and domestic consumption data to construct linear demand and supply functions. Data on bilateral trade flows are also readily available. However, information on S&H costs between regions is difficult to find. This remains a problem despite our ability to adjust S&H costs in the second stage of a numerical SPE trade model using the shadow prices on the calibration constraints in the solution to the first stage problem to adjust the S&H data, as indicated in section 4.4.

The underlying data that we use to construct forest trade models come from a variety of sources, with data from FAO (2020) constituting the primary source for forestry statistics. Supplementary data are available from the Government of Canada (2019), BC Government (2019), Random Lengths (various years), the Global Forest Products Model (GFPM) at the University of Wisconsin (available at https://buongiorno.russell.wisc.edu/gfpm/, accessed September 4, 2020), the U.S. Department of Agriculture's Forest Service (USDA Forest Service, 2019), the United Nations Economic Commission for Europe (UNECE, 2019), and van Kooten and Johnston (2014). (Data were previously available from the University of Washington's Center for International Trade in Forest Products (CINTRAFOR). For information on the CINTRAFOR model, see Perez-Garcia, 1993). Where FAO data are either unavailable, or observations are missing, supplementary data are used.

The FAO provides annual production and trade data for a number of forest products dating back to 1961. The data are collected through annual questionnaires conducted by the FAO Forestry Department in partnership with the International Tropical Timber Organization, the Statistical Office of the European Communities (Eurostat), and the UNECE. In cases where countries fail to provide information through the questionnaire, the FAO estimates production and trade of wood products through trade journals, statistical yearbooks and other sources. Where data are unavailable, the FAO repeats historical information from the previous years. Although in some instances the quality of the FAO data may be less than desired, they are nonetheless consistently available at a country level, and provide information on the destinations of various forest product exports and the origins of imports. Information on country-to-country trade flows is critical for implementing the positive mathematical programming calibration method. Since Canada and the U.S. are broken down into seven and five sub-regions, respectively, the FAO data need to be adjusted using local information (see also Chapter 5, this volume). Further, information from Canada and the U.S. is then used to reconcile missing observations in the FAO dataset.

The data analysis begins with the collection and calculation of each region's technical ability to produce logs and wood products. First, a region's ability to produce logs is a function of the annual allowable cut (AAC), which is the amount of wood permitted to be sustainably harvested, and a region's ability to convert coniferous logs into industrial roundwood. Data on AAC are available from FAO (2020), the USDA Forest Service (2019), also Howard (2001), Oswalt and Smith (2014), and the Canadian Forest Service's National Forestry Database (Government of Canada, 2019). Factors converting harvested coniferous timber into industrial roundwood can be determined by taking ratios of each region's production of roundwood to harvests. Industrial roundwood is assumed to be broken down into two sub-categories: (i) sawlogs and veneer logs; and (ii) round and split pulpwood. For both categories, the FAO provides regional production and trade flows. However, in the model in Appendix 4.C, we are not concerned to replicate pulpwood trade as there is simply too little trade of pulpwood.

The ability to recover coniferous wood products (lumber, plywood, particleboard, fiberboard, pulp and wood pellets) from their respective log inputs can be calculated as the ratio of production to inputs. The FAO differentiates coniferous from non-coniferous lumber, allowing for a simple calculation of regional coniferous lumber recovery factors. This is not the case for other wood products, however. First, plywood and veneer sheets are reported as an aggregate of coniferous and non-coniferous fiber by the FAO. Thus, to estimate regional coniferous plywood and veneer sheet production, the reported aggregate data would need to be adjusted by taking the proportion of coniferous sawlogs and veneer logs consumed in a region and multiplying by total regional production of plywood and veneer sheet. A similar adjustment would be applied to particleboard. Fiberboard and pulp are also reported as an aggregate of softwood and hardwood by the FAO. Since these products use fiber primarily from pulpwood, they can be adjusted using the reported proportion of regional coniferous pulpwood consumption multiplied by the total aggregated production of the respective product. Wood pellet data can be collected irrespective of whether pellets are produced using coniferous or non-coniferous fiber. The FAO does not currently report wood pellet statistics directly; thus, we rely on other sources (e.g. Lamers *et al.*, 2012; Government of Canada, 2019; EuroStat, 2019) and adjust regional production based on the proportion of coniferous industrial roundwood consumption by each region. Finally, a region's ability to recover chips and residuals from sawmilling is determined from various sources such as the BC Government (2019), UNECE (2019), and Government of Canada (2019).

Regional consumption of logs and wood products is based on apparent consumption (production plus imports minus exports) since the FAO only reports production and trade. For Canada and the United States, regional consumption of logs can be determined from production, while regional exports of logs can be allocated on the basis of various statistical sources (e.g. BC Government, 2019) and trade publications. Reports from Random Lengths (various years) are employed. Regional production of lumber needs to be determined first from regional production of coniferous roundwood. The forestry statistics for doing so can be found in Government of Canada (2019) and BC Government (2019) for Canada, and USDA Forest Service (2019). Canadian and U.S. population data, which are used to disaggregate Canada and the U.S. into regions, are available from Statistics Canada and the U.S. Census Bureau, while world population data can be obtained from the FAO (2020).

Regional wood product consumption in Canada and the U.S., on the other hand, can be determined by allocating total consumption across regions by their proportion of population. The same can be done for regional imports – national imports allocated across regions according to population. Exports from any Canadian or U.S. region to any other country/region in the model must be determined by allocating national exports to those countries/regions by regional production, but then making adjustments based on other sources of information.

It is important to note that, in many circumstances, bilateral trade flows of wood products are reported by the FAO as an aggregate of coniferous and non-coniferous products. Thus, the trade matrices for aggregated products need to be adjusted in a similar fashion to production. Specifically, exports from a given country are adjusted based on the proportion of coniferous inputs used in the respective region.

Data on prices can be found in the various issues of *Random Lengths* and the timber database of the UNECE (2019). The price elasticities of demand for wood products are difficult to find but can be calculated from information provided in various sources (Niquidet and Tang, 2013; Oswalt and Smith, 2014; van Kooten and Johnston, 2014; BC Government, 2019; FAO, 2020). Various elasticity estimates are also provided in the data files discussed in the Appendices to this chapter. In many cases where data are missing, supply elasticities are set to 1.0, because then the slopes of the products in question are simply given by the ratio of the base production (manufacturing) costs divided by output, while the supply function begins at the origin (intercepts equal zero). Manufacturing costs can be found in UNECE (2019) (also see van Kooten and Johnston, 2014).

The shadow prices associated with the calibration constraints in the first phase of the PMP procedure are then used to adjust the observed shipping and handling and other transaction costs; associated shadow prices may be either positive (e.g. an unobserved non-tariff cost) or negative (perhaps a subsidy not accounted for). For log shipments, the transportation costs are multiplied by 1.27 to account for the extra volume required to transport logs compared with lumber.

4.6 Grouping Regions

The task of creating bilateral trade flows among regions is exacerbated by the need to identify the nations that are to be grouped into regions. Once countries are grouped into regions, bilateral trade flows between regions (inter-regional) must be identified, while trade flows within regions (intra-regional) must be excluded. Intra-regional trade is treated the same as movement of commodities within the same jurisdiction. Sources of trade data and the allocation of information into regions are discussed in this section.

As noted in the previous section, much of the data employed in SPE forest-product trade models comes from the FAO (2020). While the FAO provides information on wood products trade, the United Nations Comtrade Database (https://comtrade.un.org/data/, accessed September 4, 2020), a pseudonym for the UN's International Trade Statistics Database, provides data on bilateral trade. Comtrade delivers data on exports of numerous commodities from one country to each of its trading partners, and, at the same time, imports by any one country from each of its trading partners. In some cases, country A may report a level of exports to country B that differs from the imports that country B reports as coming from country A. As a result, the analyst needs to reconcile these sometimes disparate data; reconciliation often requires the use of data from other sources such as the FAO, U.S. Department of Agriculture, WTO (2019), or an industry source. Despite this problem, the FAO and Comtrade data are likely the best and most consistent data that one can find.

A first step in collecting the data is to determine the harmonized system (HS) code for the product(s) of interest. For example, the international 2-digit HS codes for 'wood & wood products' range from HS44 ('wood and articles of wood') to HS49 ('printed books, newspapers, picture …') as found at https://www.foreign-trade.com/reference/hscode.htm (accessed September 4, 2020). The 4-digit HS code for sawnwood is 4407, while the 6-digit HS codes for sawnwood are 440711 (pine), 440712 (spruce and fir), and 440719 (coniferous sawnwood other than spruce-pine-fir (SPF)). The analyst might want to consider all coniferous (softwood) sawnwood. In that case, it is necessary to go to the Comtrade data and sum these three categories for each country's exports and imports. Once that is accomplished, the countries can be grouped into larger regions – this is necessary in the REPA trade model and, in some cases, in the Global Forest Products Model (see Chapter 6).

In Appendix 4.D, an R program is provided that enables one to group country-level bilateral trade data for any given commodity into region-level bilateral trade data for that commodity. Also provided in the Appendix is an example of grouping countries into regions for trade in sawnwood – the 4-digit commodity classification HS4407.

APPENDIX 4.A: GAMS Code for Base Model

The base model and the data required to run it are provided in the 'Base Lumber Trade Model' folder under the 'Trade Models' tab at the website https://www.vkooten.net/ (accessed September 4, 2020) Other models discussed in this book are also provided at that site.

GAMS Code

```
* Spatial equilibrium lumber trade model

$Oneolcom
$eolcom #
SETS s supply regions /aus, bcc, bci, al, ac, roc, chl, chn, fin, jap, nz,
        rus, swe, usn, uss, usw, rola, ROE, roa, row /
;
alias(s,d); #All regions considered to be producers and consumers

TABLE sup(s,*) supply info
        Supply      sprice     selast
*     (000s m3)  ($/000m3) elas of supply
aus     4602.7    226000   0.5
bcc     4603.0    204300   0.529
bci     16528.9   195100   0.529
al      4608.3    196100   0.415
ac      3227.6    214700   0.57
roc     9715.4    225400   0.415
chl     5862.7    211100   0.5
chn     25145.7   255000   0.5
fin     9008      227800   0.5
jap     15508.9   325000   0.5
nz      3946.6    212000   0.5
rus     23467.5   211100   0.5
swe     16173.2   248700   0.5
usn     4026.3    227200   0.2
uss     37391.7   193100   0.937
usw     16934.3   212500   0.866
rola    14429.2   205000   1
ROE     62519.5   275000   1
roa     15497.9   230000   1
row     23006.3   208900   1.5
;

TABLE dem(d,*) demand info
      demand     dprice     delast
*   (000s m3)  ($/000m3)  elas of demand
aus  5171.1     226000    -0.16
bcc  1837.4     204300    -0.16
bci  487.5      195100    -0.16
al   1923.3     196100    -0.16
ac   1198.4     214700    -0.16
roc  12087.3    225400    -0.16
chl  3777.7     211100    -0.21
```

```
chn  35406.6  255000  -0.21
fin  4345.1   227800  -0.16
jap  21763.7  325000  -0.16
nz   1720.5   212000  -0.16
rus  11310.2  211100  -0.14
swe  5633.8   248700  -0.16
usn  31142.2  227200  -0.16
uss  24578.2  193100  -0.16
usw  15963.9  212500  -0.16
rola 15308.2  205000  -0.56
ROE  77047.5  275000  -0.17
roa  16411.6  230000  -0.21
row  29089.8  208900  -0.2
;

TABLE lumflow(s,d)
$Include lumberflow.gms
;
TABLE SH(s,d)   Shipping & handling costs on lumber exported from s to d
$Include LumSH.gms
;
TABLE taxduty(s,d) Duties on lumber exported from s to d
$Include Duty.gms
;

* ------- Lumber Manufacturing Cost and Lumber Demand Parameters ------
PARAMETERS slope(s)   Slope parameter of lumber manufacturing cost
   sint(s)        Intercept parameter of manufacturing cost
   dslope(d)      Slope of lumber demand function
   dint(d)        Intercept of lumber demand function
   tax(s,d)       lumber tariffs imposed by demand region
;
* Assume elasticity of supply equals 1.0 so supply of lumber goes thru origin
slope(s) = sup(s,'sprice')/sup(s,'supply');
sint(s) = 0;
dslope(d) = -1*dem(d,'dprice')/(dem(d,'delast')*dem(d,'demand'));
dint(d) = dem(d,'dprice') + dslope(d)*dem(d,'demand');
tax(s,d) = dem(d,'dprice')*taxduty(d,s); #Converts ad valorem tax to per unit

* ----------- CONSTRUCTION OF THE MODEL EQUATIONS ---------
POSITIVE VARIABLES
    qd(d)   lumber consumption
    qs(s)   lumber supply
    x(s,d)  lumber trade from supply to demand regions
;
```

```
VARIABLE
   Z   total surplus in phase I problem ;
EQUATIONS
   obj   objective for PMP calibration phase
   ys(s)   lumber supply constraint
   yd(d)   lumber demand constraint
;

obj.. Z =E= sum(d, (dint(d)-.5*dslope(d)*qd(d))*qd(d)) - sum(s,
(sint(s)+.5*slope(s)*qs(s))*qs(s))
          - sum((s,d), SH(s,d)*x(s,d)) # multiply by 1000 to get price
          - sum(s, sum(d, x(s,d)*tax(s,d))) ;

ys(s)..   sum(d, x(s,d)) =L= qs(s);
yd(d)..   qd(d) - sum(s, x(s,d)) =L= 0;

MODEL lumbertrade /obj, ys, yd/;

OPTION qcp = cplex;
SOLVE lumbertrade using QCP maximizing Z;
* ----------------------------------------------------

* ---------- Parameters to store results ---------------
PARAMETERS lumtax(s,d) lumber taxes
   Trade(s,d)        Bilateral lumber trade
   Produced(s)       Total lumber production by region
   Consumed(d)       Total lumber consumption by region
   TotLumTax(s)      Total lumber tax paid by region $ mil
   pd(d)             Demand price of lumber $ per m3
   ps(s)             Supply price of lumber $ per m3
   CS(d)             Consumer surplus ($ mil)
   QR(s)             Producer surplus or quasi rent ($ mil)
;
 lumtax(s,d) = dem(d,'dprice')*taxduty(d,s);
 Trade(s,d) = round(x.l(s,d),2);
 Produced(s) = sum(d, x.l(s,d));
 Consumed(d) = sum(s, x.l(s,d));
 pd(d) = yd.m(d); # Also calculate as: pd(d) = dint(d) - dslope(d)*Cons(d);
 ps(s) = ys.m(s); # Also calculate as: ps(s) = sint(s) + slope(s)*Total(s);
 TotLumTax(s) = sum(d, x.l(s,d)*pd(d)*taxduty(s,d))/1000000;
 CS(d) = (.5*(dint(d)-pd(d))*qd.l(d))/1000000; # Consumer surplus
 QR(s) = (.5*ps(s)*qs.l(s))/1000000;            # Quasi rent

OPTION Trade:1:1:1; option pd:2:0:10; option pd:2:0:10;
OPTION TotLumTax:1:0:1; option CS:1:0:10; option QR:1:0:10;

DISPLAY Trade, pd, ps, TotLumTax, CS, QR;
```

```
$call gdxxrw.exe TradeResult.xls o=ExportData.gdx SQ=N @export.txt
execute_unload "TradeResult.gdx" pd, ps
* The following three lines need to appear as a SINGLE line of code in GMAS
execute 'gdxxrw.exe TradeResult.gdx O=TradeResult.xlsx SQ=N text="Lumber
    Demand Prices" rng=Price!a1 par=pd rng=Price!a2 rdim=1 text="Lumber
    Supply Prices" rng=Price!c1 par=ps rng=Price!c2 rdim=1'

execute_unload "TradeResult.gdx" Produced, Consumed
* The following three lines need to appear as a SINGLE line of code in GAMS

execute 'gdxxrw.exe TradeResult.gdx O=TradeResult.xlsx SQ=N text="Lum-
    ber Production" rng=Production!a1 par=Produced rng=Production!a2
    rdim=1 text="Lumber Consumption" rng=Production!a23 par=Con-
    sumed rng=Production!a24 rdim=1'

execute_unload "TradeResult.gdx" CS, QR
* The following three lines need to appear as a SINGLE line of code in GMAS

execute 'gdxxrw.exe TradeResult.gdx Output=TradeResult.xlsx SQ=N text="Con-
    sumer Surplus ($mil)" rng=Welfare!b1 par=CS rng=Welfare!a2 rdim=1 text-
    ="Producer Surplus ($mil)" rng=Welfare!d1 par=QR rng=Welfare!c2 rdim=1'

execute_unload "TradeResult.gdx" Trade
* The following two lines need to appear as a SINGLE line of code in GMAS
execute 'gdxxrw.exe TradeResult.gdx Output=TradeResult.xlsx SQ=N text="-
    Bilateral trade (00s m3)" rng=Tradeflow!a1 par=Trade rng=Tradeflow!a2
    rdim=1 cdim=1'
```

The following files are needed to run the above program:

LumberFlow.gms

	aus	bcc	bci	al	ac	roc	chl	chn	fin	jap	nz	rus	swe	usn	uss	usw	rola	ROE	roa	row
aus	9736.3	0.0	0.0	0.0	0.0	0.1	0.6	42.1	0.0	4.6	2.5	0.0	0.0	0.0	0.0	0.3	0.0	0.4	36.5	0.4
bcc	7.9	56.5	58.2	217.7	135.8	1369.1	0.6	472.3	0.2	274.9	4.0	0.0	0.4	769.3	607.2	394.3	2.2	45.1	15.7	23.0
bci	26.2	747.5	1663.7	781.6	407.5	2916.4	2.2	1695.9	0.7	987.0	14.5	0.0	1.4	2762.6	2180.3	1416.0	7.9	161.8	56.5	82.5
al	7.9	208.4	55.3	66.1	135.9	1370.7	0.6	472.8	0.2	275.2	4.0	0.0	0.4	770.2	607.9	394.8	2.2	45.1	15.8	23.0
ac	6.5	146.0	38.7	152.6	388.9	560.0	0.4	331.2	0.1	192.7	2.8	0.0	0.3	539.5	425.8	276.5	1.6	31.6	11.0	16.1
roc	16.6	439.4	116.4	459.4	286.6	2569.7	1.3	996.8	0.4	580.2	8.5	0.0	0.8	1623.8	1281.6	832.3	4.7	95.1	33.2	48.5
chl	17.0	0.9	0.3	1.0	0.6	6.2	4537.4	322.0	1.5	289.0	1.5	0.0	0.0	122.5	96.7	62.8	468.3	96.3	197.6	409.2
chn	0.7	0.0	0.0	0.0	0.0	0.0	0.0	17800.2	0.0	77.4	0.2	0.1	0.0	0.3	0.2	0.2	2.0	4.0	21.0	11.7
fin	10.0	0.0	0.0	0.0	0.0	0.2	0.0	74.0	4652.5	623.0	0.0	0.3	9.4	0.4	0.3	0.2	0.1	2313.5	29.8	1986.1
jap	0.0	0.0	0.0	0.0	0.0	0.0	0.0	12.6	0.0	9277.9	0.0	0.1	0.0	0.0	0.0	0.0	0.0	0.1	2.1	1.2
nz	278.0	0.0	0.0	2.0	0.0	2.0	0.1	683.0	2.4	131.0	1668.0	0.0	0.6	82.5	68.1	42.3	0.0	107.0	326.2	549.7
rus	0.1	0.0	0.0	0.0	0.0	0.0	0.0	4344.0	287.2	843.0	0.0	16889.6	11.2	0.5	0.0	1.5	0.0	2919.9	213.4	3844.5
swe	26.0	2.8	0.5	3.0	1.6	15.6	0.1	72.0	28.3	743.0	0.1	0.0	5689.3	14.3	11.3	7.3	0.1	6902.8	15.9	2868.3
usn	1.0	5.5	1.5	6.0	3.6	37.9	0.0	35.5	0.0	24.0	0.0	0.1	0.0	782.1	1318.2	686.1	51.5	5.6	4.6	6.4
uss	9.6	53.5	14.2	55.9	34.9	351.6	0.1	310.6	0.1	223.0	0.2	0.6	0.4	15510.4	3994.1	7950.1	480.7	51.6	43.0	59.5
usw	4.6	24.2	6.4	25.3	15.8	159.3	0.1	140.8	0.0	101.0	0.1	0.3	0.2	7024.5	5044.0	365.4	217.7	23.4	19.5	26.9
rola	0.3	0.4	0.1	0.4	0.3	2.7	2.2	126.6	0.0	1.6	1.1	0.2	0.1	186.5	146.4	98.1	2050.6	1.3	3.5	0.4
ROE	245.1	0.4	0.1	0.4	0.2	2.4	0.0	215.5	66.2	881.0	0.6	6.4	144.7	64.9	51.2	33.3	32.2	71846.7	6.4	156.3
roa	0.5	0.0	0.0	0.0	0.0	0.0	0.0	20.1	0.0	17.6	0.4	0.0	0.0	0.4	0.3	0.2	23.9	131.2	5546.9	4.5
row	0.3	0.0	0.0	0.0	0.0	0.1	0.0	11.0	0.1	0.9	2.1	0.0	1.4	0.3	0.2	0.1	153.4	3500.5	63.2	33594.2

LumSH.gms

	aus	bcc	bci	al	ac	roc	chl	chn	fin	jap	nz	rus	swe	usn	uss	usw	rola	ROE	roa	row
aus	0	60628	62259	63890	75800	81573	55026	43410	75660	37970	10453	70330	73740	77560	67020	58840	64791	78020	43410	53413
bcc	60628	0	9640	12840	43699	33263	51193	40630	73120	39630	55084	81770	72120	38593	31581	17489	53597	72120	40630	79760
bci	62259	9640	0	6640	40421	30108	60612	49630	82340	48320	63080	90000	81120	35457	29864	18403	60460	81120	49600	88130
al	63890	12840	6640	0	37144	26952	63190	52630	85200	52000	67100	94000	84120	32320	28148	19317	65320	84120	52530	94500
ac	75800	43699	40421	37144	0	12879	42262	82260	38450	78519	73388	46830	34450	9844	32077	46870	37712	43180	64370	58283
roc	81573	33263	30108	26952	12879	0	51816	94920	40844	90080	85360	58840	46450	5964	20961	34606	49713	54840	76370	72560
chl	55026	51193	60612	63190	42262	51816	0	49500	65390	49970	46595	68530	63470	43650	36350	40050	21450	60500	49000	68516
chn	43410	40630	49630	52630	82260	94920	49500	0	100150	8200	50481	52670	96100	94654	78519	56450	85353	97150	3000	62840
fin	75660	73120	82340	85200	38450	40844	65390	100150	0	92150	80807	8390	4020	43180	41180	65249	54830	11990	99000	50830
jap	37970	39630	48320	52000	78519	90080	49970	8200	92150	0	42882	56670	95000	88650	77600	64450	72870	96150	10172	71490
nz	10453	55084	63080	67100	73388	85360	46895	50481	80807	42882	0	78590	82510	78860	68830	66900	68307	86640	50480	57096
rus	70330	81770	90000	94000	46830	58840	68530	52670	8380	56670	78590	0	11330	48180	48180	69250	57240	15180	22150	69160
swe	73740	72120	81120	84120	34450	46450	63470	96100	4020	95000	82510	11330	0	43180	41180	64280	52990	9810	98000	50290
usn	77560	38593	35457	32320	9844	5964	43650	94654	43180	88650	78860	48180	43180	0	22828	38943	44710	48880	73024	60940
uss	67020	31581	29864	28148	32077	20961	36350	78519	43180	77600	68830	48180	41180	22828	0	22069	38274	43880	67200	46970
usw	58840	17489	18403	19317	46870	34606	40050	56450	65249	64450	66900	69250	64250	38943	22069	0	48063	67950	44820	77870
rola	64791	53597	60460	65320	37712	49713	21450	85353	54830	72870	68307	57240	52990	44710	38274	48063	0	49210	57350	45770
ROE	78020	72120	81120	84120	43180	54840	60500	97150	11990	96150	86640	15180	9810	48880	43880	67950	49210	0	98000	48160
roa	43410	40630	49600	52530	64370	76370	49000	3000	99000	10172	50480	22150	98000	73024	67200	44820	57350	98000	0	62840
row	53413	79760	88130	94500	58283	72560	68516	62840	50830	71490	57096	69160	50290	60940	46970	77870	45770	48160	62840	0

Duty.gms

	aus	bcc	bci	al	ac	roc	chl	chn	fin	jap	nz	rus	swe	usn	uss	usw	rola	ROE	roa	row
aus	0	0	0	0	0	0	0.08	0	0.01	0.02	0.05	0.1	0.01	0	0	0	0	0	0	0
bcc	0.05	0	0	0	0	0	0.08	0	0.01	0.02	0.05	0.1	0.01	0.129	0.129	0.129	0	0	0	0
bci	0.05	0	0	0	0	0	0.08	0	0.01	0.02	0.05	0.1	0.01	0.129	0.129	0.129	0	0	0	0
al	0.05	0	0	0	0	0	0.08	0	0.01	0.02	0.05	0.1	0.01	0.1875	0.1875	0.1875	0	0	0	0
ac	0.05	0	0	0	0	0	0.08	0	0.01	0.02	0.05	0.1	0.01	0	0	0	0	0	0	0
roc	0.05	0	0	0	0	0	0.08	0	0.01	0.02	0.05	0.1	0.01	0.076	0.076	0.076	0	0	0	0
chl	0.05	0	0	0	0	0	0	0	0.01	0.02	0.05	0.1	0.01	0	0	0	0	0	0	0
chn	0.05	0	0	0	0	0	0.08	0	0.01	0.02	0.05	0.1	0.01	0	0	0	0	0	0	0
fin	0.05	0	0	0	0	0	0.08	0	0	0.02	0.05	0.1	0.01	0	0	0	0	0	0	0
jap	0.05	0	0	0	0	0	0.08	0	0.01	0	0.05	0.1	0.01	0	0	0	0	0	0	0
nz	0.05	0	0	0	0	0	0.08	0	0.01	0.02	0	0.1	0.01	0	0	0	0	0	0	0
rus	0.05	0	0	0	0	0	0.08	0	0.01	0.02	0.05	0	0.01	0	0	0	0	0	0	0
swe	0.05	0	0	0	0	0	0.08	0	0.01	0.02	0.05	0.1	0	0	0	0	0	0	0	0
usn	0.05	0	0	0	0	0	0.08	0	0.01	0.02	0.05	0.1	0.01	0	0	0	0	0	0	0
uss	0.05	0	0	0	0	0	0.08	0	0.01	0.02	0.05	0.1	0.01	0	0	0	0	0	0	0
usw	0.05	0	0	0	0	0	0.08	0	0.01	0.02	0.05	0.1	0.01	0	0	0	0	0	0	0
rola	0.05	0	0	0	0	0	0.08	0	0.01	0.02	0.05	0.1	0.01	0	0	0	0	0	0	0
ROE	0.05	0	0	0	0	0	0.08	0	0.01	0.02	0.05	0.1	0.01	0	0	0	0	0	0	0
roa	0.05	0	0	0	0	0	0.08	0	0.01	0.02	0.05	0.1	0.01	0	0	0	0	0	0	0
row	0.05	0	0	0	0	0	0.08	0	0.01	0.02	0.05	0.1	0.01	0	0	0	0	0	0	0

Solution to the base model

Upon solving the GAMS model, we obtain the results in Tables 4.A.1 and 4.A.2. The first table provides the trade flows among the regions, while the second provides the domestic prices, consumption and production, and welfare measures.

Table 4.A.1. Regional trade flows projected by the base SPE trade model ('000s m^3).

	aus	bcc	bci	al	ac	roc	chl	chn	fin	jap	nz	rus	swe	usn	uss	usw	rola	ROE	roa	row
aus	4,365																			
bcc		1,826						2,959												
bci			484			0		1,508		9,548				5,690						
al				1,907		1,955								991						
ac					1,186									2,257						
roc						10,063														
chl							3,883	1,202												
chn								24,949												
fin									4,330									4,876		
jap										12,011										
nz	849									990	1,746									
rus												11,357						7,606	3,811	
swe													5,684					9,597		
usn														4,233						
uss														17,716	24,112					
usw																16,052			300	
rola																	14,981			
ROE																		56,277		
roa								4,846											12,001	
row																				27,862

Table 4.A.2. Price, production and consumption, and welfare measures: solution to the base SPE trade model.

Region	Price ($/m³)	Production ('000s m³)	Consumption ('000s m³)	Consumer surplus ($ mil)	Quasi-rent ($ mil)
Australia	214.32	4,364.84	5,213.86	3,712.73	467.74
BC Coast	212.38	4,785.04	1,825.77	1,158.27	508.12
BC Interior	203.38	17,230.35	484.19	293.20	1,752.15
Alberta	206.52	4,853.09	1,906.95	1,158.67	501.12
Atlantic Canada	228.99	3,442.46	1,185.64	787.01	394.15
Rest of Canada	233.47	10,063.18	12,018.07	8,416.74	1,174.72
Chile	183.11	5,085.35	3,882.89	2,005.96	465.59
China	253.01	24,949.43	35,464.63	21,567.39	3,156.22
Finland	232.80	9,205.70	4,329.84	3,071.48	1,071.54
Japan	251.70	12,011.03	22,549.07	23,727.83	1,511.59
New Zealand	192.57	3,584.85	1,745.73	1,173.51	345.16
Russia	204.86	22,773.78	11,357.01	8,597.81	2,332.71
Sweden	234.98	15,280.95	5,683.53	4,456.16	1,795.36
U.S. North	238.84	4,232.52	30,886.99	21,750.06	505.44
U.S. South	216.01	41,827.71	24,111.66	14,273.70	4,517.57
U.S. West	205.19	16,351.73	16,051.77	10,718.05	1,677.60
Rest of Latin America	212.83	14,980.60	14,980.60	2,683.31	1,594.19
Rest of Europe	247.54	56,276.57	78,355.42	64,451.54	6,965.34
Rest of Asia	250.01	16,846.20	16,111.76	8,661.91	2,105.86
Rest of world	252.99	27,861.89	27,861.89	13,936.67	3,524.38

APPENDIX 4.B: GAMS Code for Calibrated Base Model

The GAMS code found in the folder 'Extended Lumber Trade Model' under the 'Trade Models' tab at the website https://www.vkooten.net/ (accessed September 4, 2020).

GAMS Code

```
* Spatial equilibrium lumber trade model
$Oneolcom
$eolcom #

SETS s supply regions /aus, bcc, bci, al, ac, roc, chl, chn, fin, jap, nz,
        rus, swe, usn, uss, usw, rola, ROE, roa, row /
;
alias(s,d); #All regions considered to be producers and consumers

*=== IMPORT DATA FROM SPREADSHEET ===*
* Build gdxxrw instructions file
```

```
$onecho > import.txt
par=sup rng=Data!B2:E22 Rdim=1 Cdim=1
par=dem rng=Data!G2:J22 Rdim=1 Cdim=1
par=lumflow rng=Trade!A1:U21 Rdim=1 Cdim=1
par=SH rng=Transport!A1:U21 Rdim=1 Cdim=1
par=taxduty rng=Duty!A1:U21 Rdim=1 Cdim=1
$offecho

* Create and import GDX file with import data
$call gdxxrw.exe InputData.xlsx o=ImportData.gdx @import.txt
$gdxin ImportData.gdx

PARAMETERS
  sup(s,*)        Supply data matrix (see below)
  dem(d,*)        Demand data matrix (see below)
  taxduty(s,d)    Matrix of lumber import duties
  SH(s,d)         Shipping and handling costs ($ per m3)
  lumflow(s,d)    Lumber trade flow matrix for calibrating model (observed)

$load sup dem lumflow SH taxduty
$gdxin

* ------- Lumber Manufacturing Cost and Lumber Demand Parameters ------
PARAMETERS slope(s)    Slope parameter of lumber manufacturing cost
  sint(s)       Intercept parameter of manufacturing cost
  dslope(d)     Slope of lumber demand function
  dint(d)       Intercept of lumber demand function
  tax(s,d)      lumber tariffs imposed by demand region
;
* Assume elasticity of supply equals 1.0 so supply of lumber goes thru origin
slope(s) = sup(s,'sprice')/sup(s,'supply');
sint(s) = 0;
dslope(d) = -1*dem(d,'dprice')/(dem(d,'delast')*dem(d,'demand'));
dint(d) = dem(d,'dprice') + dslope(d)*dem(d,'demand');
tax(s,d) = dem(d,'dprice')*taxduty(d,s); #Converts ad valorem tax to per unit

* ----------------------------------------------------
*   Stage I
* ----------------------------------------------------
POSITIVE VARIABLES
   x(s,d)     lumber trade from supply to demand regions
   qd(d)      quantity demanded in each region
   qs(s)      quantity supplied in each region
;
VARIABLE
   Z      total surplus in phase I problem;
```

```
EQUATIONS
   obj1       objective for PMP calibration phase
   obj2       objective for stage II
   ys(s)      lumber supply constraint
   yd(d)      lumber demand costraint
   calibrate(s,d) marginal cost exceeds marginal revenue
;

obj1..   Z = E= sum(d, (dint(d)-.5*dslope(d)*qd(d))*qd(d))
           - sum(s, (sint(s)+.5*slope(s)*qs(s))*qs(s))
           - sum((s,d), SH(s,d)*x(s,d))
           - sum(s, sum(d, x(s,d)*tax(s,d))) ;
ys(s)..   sum(d, x(s,d)) =L= qs(s);
yd(d)..   qd(d) - sum(s, x(s,d)) =L= 0;
calibrate(s,d)..  x (s,d) =E= lumflow(s,d);

MODEL tradeCal /obj1, ys, yd, calibrate/;

OPTION QCP = CPLEX;   # Other solvers are Minos5 en Conopt
SOLVE tradeCal using QCP maximizing Z;

* ---------- Parameters to store results ---------------
PARAMETERS cal(s,d)    Shadow prices of lumber flows from PMP phase
   Trade(s,d)       Bilateral lumber trade
   Produced(s)      Total lumber production by region
   Consumed(d)      Total lumber consumption by region
   ps(s)       supply price
   pd(d)       demand price
   CS(d)       Consumer surplus ($ mil)
   QR(s)       Producer surplus or quasi rent ($ mil)
   SellerDuty(s)      Total lumber duties imposed on sellers ($ mil)
   BuyerDuty(s)       Total lumber duties paid by buyers ($ mil)
   lumtax(s,d)      Duty paid on lumber shipped from s to d
   differ(s,d)      Difference in trade flows calibrated vs observed
;
  cal(s,d) = calibrate.m(s,d);
  Trade(s,d) = round(x.l(s,d));
  pd(d) = yd.m(d); #dint(d) - dslope(d)*qd.l(d);
  ps(s) = ys.m(s); #sint(s) + slope(s)*qs.l(s);
  CS(d) = (.5*(dint(d)-pd(d))*qd.l(d))/1000000; # Consumer surplus
  QR(s) = (.5*ps(s)*qs.l(s))/1000000; # Quasi rent
  Produced(s) = sum(d, x.l(s,d));
  Consumed(d) = sum(s, x.l(s,d));

OPTION Trade:1:1:1; option pd:2:0:10; option pd:2:0:10; option
BuyerDuty:1:0:1;
option SellerDuty:1:0:1; option CS:1:0:10; option QR:1:0:10;
```

```
DISPLAY Trade, pd, Produced, Consumed, CS, QR;

* -----------------------------------------------------
* Stage II
* -----------------------------------------------------
obj2..  Z = E= sum(d, (dint(d)-.5*dslope(d)*qd(d))*qd(d))
          - sum(s, (sint(s)+.5*slope(s)*qs(s))*qs(s))
          - sum((s,d), (SH(s,d)+cal(s,d))*x(s,d))
          - sum(s, sum(d, x(s,d)*tax(s,d))) ;

MODEL tradeFin /obj2, ys, yd/;

SOLVE tradeCal using QCP maximizing Z;

  Trade(s,d) = round(x.l(s,d));
  pd(d) = dint(d) - dslope(d)*qd.l(d);
  ps(s) = sint(s) + slope(s)*qs.l(s);
  lumtax(s,d) = pd(d)*taxduty(d,s)*x.l(s,d);
  SellerDuty(s) = sum(d, lumtax(s,d))/1000000;
  BuyerDuty(d) = sum(s, lumtax(s,d))/1000000;
  Trade(s,d) = round(x.l(s,d));
  Produced(s) = sum(d, x.l(s,d));
  Consumed(d) = sum(s, x.l(s,d));
  lumtax(s,d) = pd(d)*taxduty(d,s)*x.l(s,d);
  CS(d) = (.5*(dint(d)-pd(d))*qd.l(d))/1000000; # Consumer surplus
  QR(s) = (.5*ps(s)*qs.l(s))/1000000; # Quasi rent
  differ(s,d) = Trade(s,d) - lumflow(s,d);

OPTION Trade:1:1:1; option pd:2:0:10; option pd:2:0:10; option BuyerDuty:1:0:1;
option SellerDuty:1:0:1; option CS:1:0:10; option QR:1:0:10; option dispwidth=4;

DISPLAY Trade, pd, ps, SellerDuty, BuyerDuty, CS, QR, differ;

$call gdxxrw.exe TradeResult.xls o=ExportData.gdx SQ=N @export.txt

* NOTE: Indented lines need to be programmed as a single 'execute' line in
GAMS
execute_unload "TradeResult.gdx" pd, ps
execute 'gdxxrw.exe TradeResult.gdx O=TradeResult.xls SQ=N text="Lum-
        ber Demand Prices" rng=Price!a1 par=pd rng=Price!a2 rdim=1 text
        ="Lumber Supply Prices" rng=Price!d1 par=ps rng=Price!d2 rdim=1'

execute_unload "TradeResult.gdx" Produced, Consumed
execute 'gdxxrw.exe TradeResult.gdx O=TradeResult.xls SQ=N text="Lumber
        Production" rng=Production!a1 par=Produced rng=Production!a2
        rdim=1 text="Lumber Consumption" rng=Production!a23 par=Con-
        sumed rng=Production!a24 rdim=1'
```

```
execute_unload "TradeResult.gdx" CS, QR
execute 'gdxxrw.exe TradeResult.gdx Output=TradeResult.xls SQ=N text="-
        Consumer Surplus ($mil)" rng=Welfare!b1 par=CS rng=Welfare!a2
        rdim=1 text="Producer Surplus ($mil)" rng=Welfare!d1 par=QR
        rng=Welfare!c2 rdim=1'

execute_unload "TradeResult.gdx" Trade
execute 'gdxxrw.exe TradeResult.gdx Output=TradeResult.xls SQ=N text=
        "Bilateral trade (00s m3)" rng=Tradeflow!a1 par=Trade rng=Tradeflow!a2
        rdim=1 cdim=1'
```

Discussion

There are two distinct differences between the GAMS code provided in this Appendix compared with that in Appendix 4.A. The current model includes a snippet of code that reads data from an Excel file called 'InputData.xlsx', which is available in the same folder as the GAMS code. Further, the model requires a first stage that solves for the shadow prices for the calibration constraints as discussed in section 4.2. The shadow prices are then used to revise the shipping and handling (S&H) costs matrix, which is then employed in the second stage of the model. Output will differ from that in Appendix 4.A, but will look similar. The reader is encouraged to run the above model and compare the results with those in Appendix 4.A.

APPENDIX 4.C: GAMS Code for Calibrated Model using MCP

The listing in this Appendix is for a 20-region model with eight forest products, including logs. The model consists of two stages – a first stage is used to calibrate the model, followed by a second stage that solves the calibrated model. Again the GAMS code and Excel file with the underlying data are found in the folder 'Multiple regions & products' under the 'Trade Models' tab at the website https://www.vkooten.net/ (accessed September 4, 2020).

GAMS Code

```
$TITLE Multi-Product Forest Trade Model
* This model consists of 20 regions and seven products

$Oneolcom
$eolcom #
```

```
$OFFSYMLIST OFFSYMXREF
OPTION LIMROW = 0
OPTION LIMCOL = 0
OPTION RESLIM = 50000
OPTION ITERLIM = 1000000

OPTION LP=CPLEX
OPTION QCP=MINOS5
OPTION NLP=MINOS5
OPTION DNLP=SNOPT

SETS
r     region names
p     all products
s(r)  product supply regions
d(r)  product demand regions
;
alias(r,rr);

* --- IMPORT DATA FROM SPREADSHEET ----*
* Build gdxxrw instructions file
$onecho > import.txt
par=produce rng=Production!B4:L24 Rdim=1 Cdim=1
par=consume rng=Consumption!B4:L24 Rdim=1 Cdim=1
par=price rng=Prices!B4:L24 Rdim=1 Cdim=1
par=manucost rng=ManuCost!B4:L24 Rdim=1 Cdim=1
par=elastic rng=Elastic!B4:L24 Rdim=1 Cdim=1
par=recov rng=recov!B4:L24 Rdim=1 Cdim=1
par=data rng=data!B2:D22 Rdim=1 Cdim=1
par=lumtax rng=lumtax!A1:U21 Rdim=1 Cdim=1
par=logtax rng=logtax!A1:U21 Rdim=1 Cdim=1
par=trans rng=trans!A1:U21 Rdim=1 Cdim=1
par=lumcal rng=lumberflow!B2:V22 Rdim=1 Cdim=1
par=logcal rng=logflow!B2:V22 Rdim=1 Cdim=1
par=pulplogcal rng=pulplogflow!B2:V22 Rdim=1 Cdim=1
par=pulpcal rng=pulpflow!B2:V22 Rdim=1 Cdim=1
par=residcal rng=residflow!B2:V22 Rdim=1 Cdim=1
par=plycal rng=plyflow!B2:V22 Rdim=1 Cdim=1
par=pbcal rng=osbflow!B2:V22 Rdim=1 Cdim=1
par=fbcal rng=mdfflow!B2:V22 Rdim=1 Cdim=1
set=r rng=lumtax!A2:A21 Rdim=1 Cdim=0
set=p rng=production!D4:M4 Rdim=0 Cdim=1
set=s rng=lumtax!A2:A21 Rdim=1 Cdim=0
set=d rng=lumtax!A2:A21 Rdim=1 Cdim=0
$offecho

* Create and import GDX file with import data
$call gdxxrw.exe TradeData.xlsx o=ImportData.gdx @import.txt
$gdxin ImportData.gdx
```

```
PARAMETERS
  produce(r,p)        Data matrix (see below)
  consume(r,p)        Data matrix (see below)
  price(r,p)          Data matrix (see below)
  manucost(r,p)       Data matrix (see below)
  elastic(r,p)        Data matrix (see below)
  data(r,*)           Data matrix (see below)
  recov(r,p)          Data matrix (see below)
  lumtax(r,rr)        Matrix of lumber import duties
  logtax(r,rr)        Matrix of log export taxes
  trans(r,rr)         Transportation costs
  lumcal(r,rr)        Lumber trade flow matrix for calibrating model
  logcal(r,rr)        Log trade flow matrix for calibrating model
  pulplogcal(r,rr)    pulp trade flow matrix for calibrating model
  pulpcal(r,rr)       pulplog trade flow matrix for calibrating model
  residcal(r,rr)      residual trade flow matrix for calibrating model
  plycal(r,rr)        plywood & veneer trade flow matrix for calibrating model
  pbcal(r,rr)         osb trade flow matrix for calibrating model
  fbcal(r,rr)         mdf trade flow matrix for calibrating model

$load r p s d data produce consume price manucost elastic recov lumtax logtax
trans lumcal logcal pulplogcal pulpcal residcal plycal pbcal fbcal
$gdxin

SETS
  dn(p) downstream products /lum, ply, pulp, pb, fb, pell/
  mc(dn) Products whose MC shifts as Ind RW prices change /lum, ply, pulp/
;

* ----------- Recovery calculations ----------------------------------
SCALARS
  recov_chips_lumber     Pulp chip recovery from lumber manufact /0.35/
  recov_chips_plywood    Pulp chip recovery from plywood manufact /0.20/
  recov_residuals        Bark planer shavings & sawdust recovery sawmill /0.21/
  recov_pb               Particleboard recovery from fibre /0.79/
  recov_fb               Particleboard recovery from fibre /0.86/
;

PARAMETER Impact(dn) Impact of shift variable on MC curves
/lum     0.7
ply      0.6
pulp     0.5/;

PARAMETERS
  bs(r,p)   Slope of MC curve
  as(r,p)   Intercept of MC curve
  bd(r,p)   Slope of demand curve
  ad(r,p)   Intercept of demand curve
;
```

```
* -------------- MC and Demand curve parameters ---------------
bs(r,p) = manucost(r,p)/produce(r,p);
as(r,p) = 0;
bd(r,p) = -1*price(r,p)/(elastic(r,p)*consume(r,p));
ad(r,p) = price(r,p) + bd(r,p)*consume(r,p);
* ------- Log and Lumber Tax Parameters and Misc ---------------
PARAMETERS
  ytax(r,rr)         lumber taxes
  xtax(r,rr)         log taxes that later change in the analysis
  shadow(r,rr,p)        Shadow prices from PMP constraints
;

ytax(r,rr) = price(r,'lum')*lumtax(r,rr);
xtax(r,rr) = price(r,'saw')*logtax(r,rr);

* ---------- CONSTRUCTION OF THE MODEL EQUATIONS -------------------------- *
POSITIVE VARIABLES
* Pulp and pulpwood variables
  pw_pulp(r)      pulpwood going to pulp
  pw_other(r)       pulpwood going to other (panels)
  chips_pulp(r)
* flow variables
  chips(r)           pulp chips produced from sawmilling and plywood
  residuals(r)          bark planer shavings sawdust from sawmills & plywood
* Residual variables
  fibre_pb(r)
  fibre_fb(r)
  fibre_pellet(r)  fibre going to make pellets
  ind_rw(r)             proportion to sawlogs & pulpwood cannot exceed 1
  resid_flow(r,rr)    residual & chip trade from supply to demand regions
* ---------------
  qs(r,p)            supply quantity
  qd(r,p)            demand quantity
  flow(r,rr,p)       product trade flow matrix
  zs(r,dn)           downstream product MC shifter based on log input price
;

VARIABLES
Z1      total surplus in phase I problem
Z2      total surplus with calibrated objective function
;

* -------------------------------------------------------
*       DEFINE EQUATIONS
* -------------------------------------------------------
EQUATIONS
  obj1               objective for PMP calibration phase
  obj2               revised objective after calibration
```

```
   supply(rr,p)
   demand(r,p)
   shift(r,mc)
* ---------------- Logs
   mai(r)           MAI constraint for sustainable harvests
   log_supply1(r) sawlog supply constraint for total production
   log_supply2(r) sawlog supply constraint for recovery from ind rw
   log_demand(rr)
   pmpsaw(r,rr)   calibration constraint for sawlog trade flows
* ---------------- Lumber
   lum_prod(r)    sawnwood production constraint
   pmplum(r,rr)   calibration constraint for lumber trade flows
* ---------------- Chips and residuals
   chips_prod(r)       Pulp chips produced from sawmilling
   residuals_prod(r)   residuals bark planer shavings sawdust from sawmills
* ---------------- Plywood
   ply_prod(r)    Plywood production constraint
   pmpply(r,rr)   calibration constraint for plywood trade flows
* ---------------- Particleboard
   pb_prod(r)     Particleboard production constraint
   pmppb(r,rr)    calibration constraint for Particleboard trade flows
* ---------------- Fibreboard
   fb_prod(r)     Fibreboard production constraint
   pmpfb(r,rr)    calibration constraint for Fibreboard trade flows
* ---------------- pulp
   pulp_prod(r)   pulp production constraint
   pmppulp(r,rr)  calibration constraint for pulp trade flows
* ---------------- pulpwood
   pw_prod(r)     pulpwood production constraint
   pw_supply(r)   pulpwood recovery from ind rw
* ---------------- pellets
   pellet_prod1(r)  pulp production constraint
   pellet_prod2(r)  pulp production constraint
* ---------------- Fibre flow
   Fibre(r)       pulpwood and pulp chips fibre flow constraint
   residual_flow(r) residual trade flow
   log_sum(r)   sum of sawlog and pulpwood must not exceed ind rw
;

* --------------------------------------
*     MODEL EQUATIONS
* --------------------------------------

obj1.. Z1 =E= sum((r,dn), ad(r,dn)*qd(r,dn)-.5*bd(r,dn)*qd(r,dn)*qd(r,dn))
        - sum((r,dn), (as(r,dn)+impact(dn)*zs(r,dn))*qs(r,dn)
                +.5*bs(r,dn)*qs(r,dn)*qs(r,dn))
```

```
         - sum((r,rr,dn), trans(r,rr)*flow(r,rr,dn)) - sum(r, 0.5*bs(r,'log')*qs(r,'log')*
         qs(r,'log'))
         - sum((r,rr), 1.27*trans(r,rr)*flow(r,rr,'saw')) - sum(r, sum(rr, flow(r,rr,'lum')*
         ytax(r,rr)))
         - sum(rr, sum(r, flow(r,rr,'saw')*xtax(r,rr))) - sum((r,rr), trans(r,rr)*resid_flow(r,rr)) ;

obj2..   Z2 = E= sum((r,dn), ad(r,dn)*qd(r,dn)-.5*bd(r,dn)*qd(r,dn)*qd(r,dn))
         - sum((r,dn), (as(r,dn)+impact(dn)*zs(r,dn))*qs(r,dn)
                   +.5*bs(r,dn)*qs(r,dn)*qs(r,dn))
         - sum((r,rr,dn), (shadow(r,rr,dn) + trans(r,rr))*flow(r,rr,dn))
         - sum(r, 0.5*bs(r,'log')*qs(r,'log')*qs(r,'log')) - sum((r,rr), (shadow(r,rr,'saw')
         + 1.27*trans(r,rr))*flow(r,rr,'saw')) - sum(r, sum(rr, flow(r,rr,'lum')*ytax(r,rr)))
         - sum(rr, sum(r, flow(r,rr,'saw')*xtax(r,rr))) - sum((r,rr), trans(r,rr)*resid_
         flow(r,rr)) ;

* ------ MODEL CONSTRAINTS ----------------------
* ------- Logs -----------------------------
mai(r)..              qs(r,'log') =L= data(r,'aac')*recov(r,'log')*1.01;
log_supply1(r)..      sum(rr, flow(r,rr,'saw')) =L= qs(r,'saw');
log_supply2(r)..      qs(r,'saw') =L= qs(r,'log')*recov(r,'saw');
log_demand(rr)..      sum(r, flow(r,rr,'saw')) =G= qd(rr,'saw');
pmpsaw(r,rr)..        flow(r,rr,'saw') =E= logcal(r,rr);
shift(r,mc)..         zs(r,mc) =E= as(r,'log') + bs(r,'log')*qs(r,'log');
* -------- Downstream product constraints ---------------
supply(r,dn)..        sum(rr, flow(r,rr,dn)) =L= qs(r,dn);
demand(rr,dn)..       sum(r, flow(r,rr,dn)) =G= qd(rr,dn);
* ----------- Lumber constraints -----------------------
lum_prod(rr)..        qs(rr,'lum') =L= recov(rr,'lum')*qd(rr,'saw');
pmplum(r,rr)..        flow(r,rr,'lum') =E= lumcal(r,rr);
* ---------- Plywood constraints ----------------------
ply_prod(rr)..        qs(rr,'ply') =L= recov(rr,'ply')*qd(rr,'saw');
pmpply(r,rr)..        flow(r,rr,'ply') =E= plycal(r,rr);
* -------- Particleboard constraints ------------------
pb_prod(r)..          qs(r,'pb') =L= recov(r,'pb')*fibre_pb(r);
pmppb(r,rr)..         flow(r,rr,'pb') =E= pbcal(r,rr);
* ------- Fibreboard constraints ---------------------
fb_prod(r)..          qs(r,'fb') =L= recov(r,'fb')*fibre_fb(r);
pmpfb(r,rr)..         flow(r,rr,'fb') =E= fbcal(r,rr);
* -------- Pulp constraints -------------------------
pulp_prod(r)..        qs(r,'pulp') =L= 0.4*(pw_pulp(r)+ chips_pulp(r));

$OnText # With this command the following lines are not read by the solver
0.4 converts m3 roundwood to mt of pulp; Other conversions 2.5 to 6.0 mt
pulp per m3
https://www.forestresearch.gov.uk/tools-and-resources/statistics/forestry-statistics/
forestry-statistics-2016-introduction/sources/timber/conversion-factors/ (accessed
September 4, 2020)
```

```
$OffText

pmppulp(r,rr)..  flow(r,rr,'pulp') =E= pulpcal(r,rr);
* ------- Pulpwood constraints ----------------------
pw_prod(r)..     pw_pulp(r) + pw_other(r) =L= qs(r,'pw');
pw_supply(r)..   qs(r,'pw') =L= qs(r,'log')*recov(r,'pw');
* -------- Pellet constraints ------------------------
pellet_prod1(r)..  qs(r,'pell') =L= 0.65*0.667*fibre_pellet(r);
* See above reference: 0.65 converts m3 roundwood to m3 of pellet;
* 0.667 converts mt to m3 of pellet (650 kg per m3)
pellet_prod2(r)..  qs(r,'pell') =L= data(r,'PelCap');
* -------- Material flow constraints -------------------
chips_prod(r)..  chips(r) =E= qs(r,'lum')*recov_chips_lumber +
qs(r,'ply')*recov_chips_plywood;
* chips produced depend on chips from sawmills & from plywood manufacturing
residuals_prod(r)..  residuals(r) =E= qs(r,'lum')*recov_residuals;
* Residuals produced are from sawmilling
fibre(rr).. fibre_pb(rr) + fibre_fb(rr) + fibre_pellet(rr) + chips_pulp(rr) =L= sum(r,
resid_flow(r,rr));
* pulp and residual fibre goes to particleboard, fibreboard and pulp
residual_flow(r)..  sum(rr, resid_flow(r,rr)) =L= chips(r) + residuals(r) +
pw_other(r);
log_sum(r)..  qs(r,'saw') + qs(r,'pw') =L= qs(r,'log');
* -------- End of model Constraints ---------------------------

MODEL tradePMP /obj1, mai, log_supply1, log_supply2, log_demand, supply,
      demand, lum_prod, ply_prod, pb_prod, fb_prod, chips_prod, residuals_
      prod, fibre, shift, residual_flow, pulp_prod, pw_prod, pw_supply, pellet_
      prod1, pellet_prod2, log_sum, pmpsaw, pmplum, pmpply, pmppb,
      pmpfb, pmppulp / ;

* -------------- Calibration phase ------------------------
SOLVE tradePMP using QCP maximizing Z1;
* ------------ Store PMP shadow prices -------------------
shadow(r,rr,'log') = 0;
shadow(r,rr,'saw') = pmpsaw.m(r,rr);
shadow(r,rr,'lum') = pmplum.m(r,rr);
shadow(r,rr,'ply') = pmpply.m(r,rr);
shadow(r,rr,'pb') = pmppb.m(r,rr);
shadow(r,rr,'fb') = pmpfb.m(r,rr);
shadow(r,rr,'pulp') = pmppulp.m(r,rr);
shadow(r,rr,'pell') = 0;
* --------- CALIBRATED MODEL -----------------------------
MODEL trade /obj2, mai, log_supply1, log_supply2, log_demand, supply,
      demand, lum_prod, ply_prod, pb_prod, fb_prod, chips_prod, resid-
      uals_prod, fibre, shift, residual_flow, pulp_prod, pw_prod, pw_supply,
      pellet_prod1, pellet_prod2, log_sum/ ;
SOLVE trade using QCP maximizing Z2;
```

```
VARIABLES
  log_flow(r,rr)
  lum_flow(r,rr)
  ply_flow(r,rr)
  pb_flow(r,rr)
  fb_flow(r,rr)
  pulp_flow(r,rr)
  pellet_flow(r,rr) ;

VARIABLES PS(*,*), PD(*,*), QR(*,*), CS(*,*), SR(*,*);
   PS.m(r,p) = 0.01;
   PD.m(r,p) = 0.01;
   QR.m(r,p) = 0.01;
   CS.m(r,p) = 0.01;
   SR.m(r,p) = 0.01;

   log_flow.m(r,rr) = 0.01;
   lum_flow.m(r,rr) = 0.01;
   ply_flow.m(r,rr) = 0.01;
   pb_flow.m(r,rr) = 0.01;
   fb_flow.m(r,rr) = 0.01;
   pulp_flow.m(r,rr) = 0.01;
   pellet_flow.m(r,rr) = 0.01;

   log_flow.l(r,rr) = flow.l(r,rr,'saw');
   lum_flow.l(r,rr) = flow.l(r,rr,'lum');
   ply_flow.l(r,rr) = flow.l(r,rr,'ply');
   pb_flow.l(r,rr) = flow.l(r,rr,'pb');
   fb_flow.l(r,rr) = flow.l(r,rr,'fb');
   pulp_flow.l(r,rr) = flow.l(r,rr,'pulp');
   pellet_flow.l(r,rr) = flow.l(r,rr,'pell');
   qd.l(r,'log') = qs.l(r,'log');
   qd.l(r,'pw') = qs.l(r,'pw');

loop(p,
   PS.l(r,p) = 0.001*(as(r,p) + bs(r,p)*qs.l(r,p));
   PD.l(r,p) = 0.001*(ad(r,p) - bd(r,p)*qd.l(r,p));
   QR.l(r,p) = 0.000001*0.5*bs(r,p)*qs.l(r,p)**2;
   CS.l(r,p) = 0.000001*0.5*bd(r,p)*qd.l(r,p)**2;
   SR.l(r,p) = 0.000001*supply.m(r,p)*qs.l(r,p); );
SR.l(r,'log') = 0.000001*mai.m(r)*qs.l(r,'log');

DISPLAY shadow, zs.l, impact, PS.l, PD.l, QR.l, CS.l, r, p, dn, as, bs, ad, bd;
DISPLAY qs.l, qd.l, flow.l, resid_flow.l, pw_pulp.l, pw_other.l, chips.l, residu-
als.l, fibre_pb.l, fibre_fb.l;
```

```
execute_unload "Output.gdx" shadow
execute 'gdxxrw.exe Output.gdx Output=Calibrate.xlsx SQ=N par=shadow
rng=shadow!a1'

execute_unload "Results.gdx" qd, qs, PS, PD, QR, CS, SR
execute 'gdxxrw.exe Results.gdx Output=Result01.xlsx SQ=N var=qs.l
        rng=production!a1 var=qd.l rng=consumption!a1 var=PS.l rng=S-
        price!a1 var=PD.l rng=Dprice!a1 text="Quasi-Rent" rng=Welfare!a1
        var=QR.l rng=Welfare!a2 text="Consumer Surplus" rng=Welfare!a24
        var=CS.l rng=Welfare!a25 text="Scarcity Rent" rng=Welfare!a47
        var=SR.l rng=Welfare!a48'

execute_unload "Result.gdx" log_flow, lum_flow, ply_flow, pb_flow, fb_flow,
        pulp_flow, pellet_flow, resid_flow

execute 'gdxxrw.exe Result.gdx Output=Result02.xlsx text="Inter-regional
        Sawlog Trade Flows" SQ=N rng=log_flow!a1 var=log_flow.l rng=log_
        flow!a2 rdim=1 cdim=1 text="Inter-regional Lumber Trade Flows"
        SQ=N rng=lum_flow!a1 var=lum_flow.l rng=lum_flow!a2 rdim=1
        cdim=1 text="Inter-regional Plywood + Veneer Trade Flows" SQ=N
        rng=ply_flow!a1 var=ply_flow.l rng=ply_flow!a2 rdim=1 cdim=1 text=
        "Inter-regional Particleboard Trade Flows" SQ=N rng=pb_flow!a1
        var=pb_flow.l rng=pb_flow!a2 rdim=1 cdim=1 text="Inter-regional
        Fibreboard Trade Flows" SQ=N rng=fb_flow!a1 var=fb_flow.l rng=fb_
        flow!a2 rdim=1 cdim=1 text="Inter-regional Pulp Trade Flows" SQ=N
        rng=pulp_flow!a1 var=pulp_flow.l rng=pulp_flow!a2 rdim=1 cdim=1
        text="Inter-regional Pellet Trade Flows" SQ=N rng=pellet_flow!a1
        var=pellet_flow.l rng=pellet_flow!a2 rdim=1 cdim=1 text="Inter-re-
        gional Residual Trade Flows" SQ=N rng=resid_flow!a1 var=resid_
        flow.l rng=resid_flow!a2 rdim=1 cdim=1'
```

Discussion

The model found in this Appendix is quite different from that in Appendices 4.A and 4.B, because the model covers eight products – sawlogs, lumber, plywood, pulp, particle board (oriented strand board, OSB), fiberboard (medium density fiberboard, MDF), wood pellets, and residuals – and 20 regions and/or countries (including five Canadian and three U.S. regions). The data to run the model are found in the file 'TradeData.xlsx'. The bilateral trade matrices are provided in the Excel file, but need to be updated; to update them and other data required in the model, a user only has to make changes to the input file. The matrices describing trade flows are important because they are used to calibrate the trade model. However, at this time, there may be too little information on the country-to-country trade flows of wood pellets to be included in the calibration component of the model. The GAMS results are then reported in two files, denoted 'Results01.xlsx' and 'Results02.xlsx', with production, consumption and the various welfare measured reported in the former and trade flows in the latter.

Appendix 4.D: R code

In this Appendix, we provide the R code for allocating countries into regions. It calls data from Comtrade, which was downloaded into an Excel CSV file format and named 'comtrade.csv'. The Comtrade file was then edited to remove columns that were empty. An example of the first 19 lines of the comtrade.csv file is provided in Fig. 4.D.1, which enables the user to better understand the R call to this file. Nonetheless, the user might need to modify the R code to suite their own needs.

The file in Fig. 4.D.1 is for HS 440710, defined as 'Wood and articles of wood; wood charcoal. Wood sawn or chipped lengthwise, sliced or peeled, whether or not planed, sanded or end-jointed, of a thickness exceeding 6 mm. Coniferous'. For wood product trade (HS440710) in 2016, it includes the exports to and imports from each 'reporter country'; the user will need to filter the data by imports or exports to obtain bilateral trade flows. An example of the type of output that one can expect from the R code below is found in Table 4.D.1.

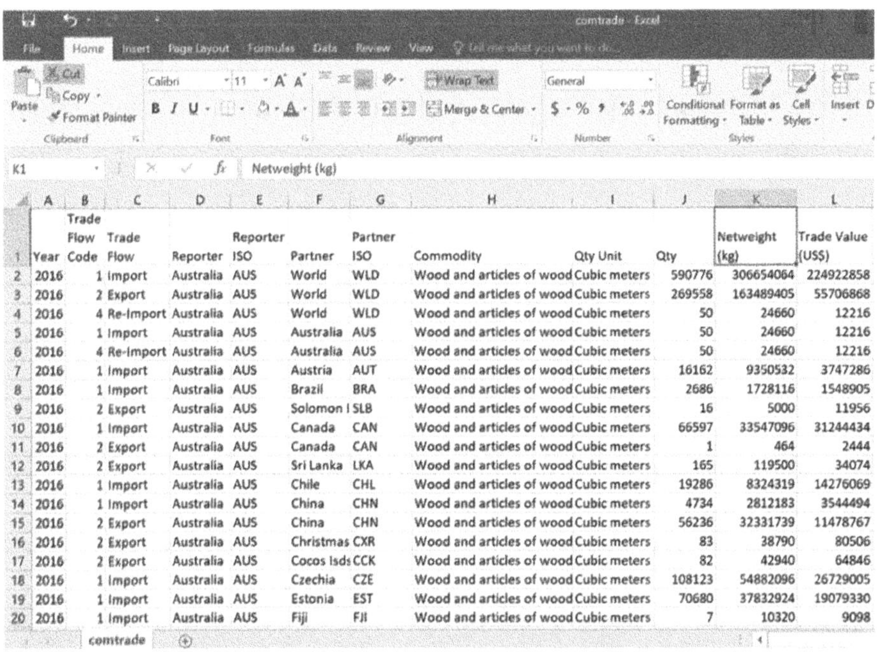

Year	Trade Flow Code	Trade Flow	Reporter	Reporter ISO	Partner	Partner ISO	Commodity	Qty Unit	Qty	Netweight (kg)	Trade Value (US$)
2016	1	Import	Australia	AUS	World	WLD	Wood and articles of wood	Cubic meters	590776	306654064	224922858
2016	2	Export	Australia	AUS	World	WLD	Wood and articles of wood	Cubic meters	269558	163489405	55706868
2016	4	Re-Import	Australia	AUS	World	WLD	Wood and articles of wood	Cubic meters	50	24660	12216
2016	1	Import	Australia	AUS	Australia	AUS	Wood and articles of wood	Cubic meters	50	24660	12216
2016	4	Re-Import	Australia	AUS	Australia	AUS	Wood and articles of wood	Cubic meters	50	24660	12216
2016	1	Import	Australia	AUS	Austria	AUT	Wood and articles of wood	Cubic meters	16162	9350532	3747286
2016	1	Import	Australia	AUS	Brazil	BRA	Wood and articles of wood	Cubic meters	2686	1728116	1548905
2016	2	Export	Australia	AUS	Solomon I	SLB	Wood and articles of wood	Cubic meters	16	5000	11956
2016	1	Import	Australia	AUS	Canada	CAN	Wood and articles of wood	Cubic meters	66597	33547096	31244434
2016	2	Export	Australia	AUS	Canada	CAN	Wood and articles of wood	Cubic meters	1	464	2444
2016	2	Export	Australia	AUS	Sri Lanka	LKA	Wood and articles of wood	Cubic meters	165	119500	34074
2016	1	Import	Australia	AUS	Chile	CHL	Wood and articles of wood	Cubic meters	19286	8324319	14276069
2016	1	Import	Australia	AUS	China	CHN	Wood and articles of wood	Cubic meters	4734	2812183	3544494
2016	2	Export	Australia	AUS	China	CHN	Wood and articles of wood	Cubic meters	56236	32331739	11478767
2016	2	Export	Australia	AUS	Christmas	CXR	Wood and articles of wood	Cubic meters	83	38790	80506
2016	2	Export	Australia	AUS	Cocos Isds	CCK	Wood and articles of wood	Cubic meters	82	42940	64846
2016	1	Import	Australia	AUS	Czechia	CZE	Wood and articles of wood	Cubic meters	108123	54882096	26729005
2016	1	Import	Australia	AUS	Estonia	EST	Wood and articles of wood	Cubic meters	70680	37832924	19079330
2016	1	Import	Australia	AUS	Fiji	FJI	Wood and articles of wood	Cubic meters	7	10320	9098

Fig. 4.D.1. Example of the modified CSV file downloaded from Comtrade (https://comtrade.un.org/data/) for Commodity HS4407

Table 4.D.1. Sample output from R file allocating countries into regions.

	AUS	CHL	CHN	USA	FIN	JPN	NZL	RUS	SWE	CAN	ROA	ROE	ROLA	ROW
AUS	-	-	56.2	0.1	-	2.8	0.6	-	-	0.0	104.7	0.0	-	18.6
CHL	22.0	-	735.5	358.6	-	256.1	5.2	-	-	3.2	745.0	51.2	647.3	450.9
CHN	1.5	-	-	0.1	-	219.1	0.1	-	-	0.1	108.5	1.9	3.0	5.5
USA	4.2	2.4	696.4	-	-	243.9	0.7	0.6	0.1	875.4	282.9	108.5	1,497.0	73.3
FIN	21.5	-	1,061.7	3.8	-	941.5	0.1	0.5	50.8	0.1	134.4	3,236.1	1.3	3,134.0
JPN	0.0	-	43.5	3.4	-	-	-	-	-	-	32.1	0.1	0.0	0.4
NZL	156.9	-	385.2	206.9	2.9	36.0	-	-	-	0.3	602.9	80.3	-	176.4
RUS	7.3	-	13,398.3	24.8	478.0	905.0	0.1	-	8.3	-	3,836.3	4,022.7	4.6	2,731.9
SWE	42.9	-	733.9	250.1	16.5	778.4	-	-	-	3.2	233.9	7,820.1	3.6	2,886.8
CAN	65.6	4.8	5,949.6	35,263.0	0.1	2,288.1	31.0	-	1.7	-	875.8	161.3	45.8	226.6

R-Code

```
# We thank Hugh Scorah for helping to write this code
# Creation of Bilateral Trade Flows
# The following code allocates countries to regions and determines the inter-
regional trade flows.

library(tidyverse)
# if you do not have tidyverse, use the following command
#install.packages("tidyverse")

# ----------------- Country listed by 3-digit ISO letters -------------------------------
# Major exporting countries
core <- c('AUS', 'CHL', 'CHN', 'USA', 'FIN', 'JPN', 'NZL', 'RUS', 'SWE', 'CAN')

output_regions <- c(core,"ROA","ROE","ROLA","ROW")

rest_asia <- c('AFG', 'BGD', 'BTN', 'HKG', 'IDN', 'IND', 'KAZ', 'KGZ', 'KHM',
   'KOR', 'LAO', 'LKA', 'MAC', 'MDV', 'MMR', 'MNG', 'MYS', 'NPL', 'PAK',
   'PHL', 'PRK', 'SGP', 'THA', 'TJK', 'TKM', 'TLS', 'UZB', 'VNM', 'OAS')
rest_europe <- c('ALB', 'ARM', 'AUT', 'AZE', 'BEL', 'BGR', 'BIH', 'BLR', 'CHE',
'CYP', 'CZE', 'DEU', 'DNK', 'ESP', 'EST', 'FRA', 'FRO', 'GBR', 'GEO', 'GRC',
'GRL', 'HRV', 'HUN', 'IRL', 'ISL', 'ITA', 'LTU', 'LUX', 'LVA', 'MDA', 'MLT', 'NLD',
'NOR', 'POL', 'PRT', 'ROU', 'SRB', 'SVK', 'SVN', 'TUR', 'UKR')
rest_latin_am <- c('ABW', 'AIA', 'ARG', 'ATG', 'BHS', 'BLZ', 'BMU', 'BRA',
'BRB', 'COL', 'CPV', 'CRI', 'CUB', 'CUW', 'CYM', 'DMA', 'DOM', 'ECU',
'GTM', 'GUY', 'GRD', 'HND', 'HTI', 'JAM', 'KNA', 'LCA', 'MEX', 'MSR', 'NIC',
'PAN', 'PER', 'SLV', 'TTO', 'URY', 'VCT', 'VEN', 'VGB')
rest_world <- c('AGO', 'ARE', 'ASM', 'ATF', 'BDI', 'BHR', 'BLM', 'CCK', 'CIV',
   'CMR', 'COD', 'COG', 'COK', 'CXR', 'DJI', 'DZA', 'EGY', 'ETH', 'FJI', 'FSM',
   'GAB', 'GHA', 'GIN', 'GMB', 'GNQ', 'IRN', 'IRQ', 'ISR', 'JOR', 'KEN',
   'KIR', 'KWT', 'LBN', 'MAR', 'MDG', 'MHL', 'MNP', 'MOZ', 'MRT', 'MUS',
   'NCL', 'NFK', 'NAM', 'NIU', 'NRU', 'OMN', 'PCN', 'PLW', 'PNG', 'PSE',
   'PYF', 'QAT', 'SAU', 'SDN', 'SEN', 'SHN', 'SLB', 'SLE', 'SOM', 'SPM', 'SSD',
   'STP', 'SUR', 'SWZ', 'SXM', 'SYC', 'SYR', 'TCA', 'TON', 'TUN', 'TUV', 'TZA',
   'UMI', 'VUT', 'WLF', 'WSM', 'YEM', 'ZAF', 'ZMB', 'ZWE')
EU28 <- c('AUT', 'BEL', 'BGR', 'CHE', 'CYP', 'DEU', 'DNK', 'ESP', 'EST', 'FIN',
   'FRA', 'GBR' 'GRC', 'GRL', 'HRV', 'HUN', 'IRL', 'ITA', 'LTU', 'LUX', 'LVA',
   'NLD', 'POL', 'PRT', 'ROU', 'SRB', 'SVK', 'SVN', 'SWE')
EU15 <- c('AUT', 'BEL', 'DEU', 'DNK', 'ESP', 'FIN', 'FRA', 'GBR', 'GRC', 'IRL',
         'ITA', 'LUX', 'NLD', 'PRT', 'SWE')
```

```
# --------------------------------COMTRADE------------------------------------------------------
##functions##
trade_flow_sort<- function(trade_frame,core,output_regions,rest_asia,rest_
europe, rest_latin_am, rest_world){
          return_matrix <- data.
          frame(matrix(0,length(core),length(output_regions)))
          rownames(return_matrix) <- core
          colnames(return_matrix) <- output_regions
for (i in 1:nrow(return_matrix)){
          current_reporter <- rownames(return_matrix)[i]
          c_table <- filter(trade_frame,reporter==current_reporter)

# need to test for null table.
  if (nrow(c_table) > 0) {
    for (j in 1:nrow(c_table)){
        partner <- c_table$partner[j]
        #print (partner)
        qty <- c_table$qty[j]
        if (partner %in% core) {return_matrix[current_reporter,partner] <- qty
        }
    else if (partner %in% rest_asia) {
        return_matrix[current_reporter,"ROA"]<-return_matrix[current_reporter,
        "ROA"] + qty
        }
    else if (partner %in% rest_europe) {
        return_matrix[current_reporter,"ROE"]<-return_matrix[current_reporter,
        "ROE"] + qty
        }
    else if (partner %in% rest_latin_am) {
        return_matrix[current_reporter,"ROLA"]     <-     return_matrix[current_
        reporter,"ROLA"] + qty
        }
    else if (partner %in% rest_world) {
        return_matrix[current_reporter,"ROW"]<-return_matrix[current_reporter,
        "ROW"] + qty
        }
    }
  }
}
return (return_matrix)
}
#---------------------------------------------------------------------------------------------
comtrade_raw <- read_csv("comtrade.csv")
comtrade <- select(comtrade_raw,reporter = 'Reporter ISO',partner = 'Partner
ISO',trade_flow = 'Trade Flow Code', qty = 'Qty',value = 'Trade Value (US$)')
```

```
#trade_flow variable takes value '1' for Import, '2' for Export, '3' for Re-export,
'4' for Re-import

####EXPORT####
comtrade_ex <- filter(comtrade,trade_flow==2)
exports <- trade_flow_sort(comtrade_ex,core,output_regions,rest_asia,rest_
      europe, rest_latin_am, rest_world)

write.csv(exports, 'int_exports.csv') # write bilateral trade exports to file

#####RE-EXPORT####

comtrade_reex <- filter(comtrade,trade_flow==3 )
reexports <- trade_flow_sort(comtrade_reex,core,output_regions,rest_
   asia,rest_europe, rest_latin_am, rest_world)

write.csv(exports, 'int_exp_rexp.csv') # write exports plus re-exports to file

####IMPORT####
comtrade_im <- filter(comtrade,trade_flow==1)
imports <- trade_flow_sort(comtrade_im,core,output_regions,rest_asia,rest_
   europe, rest_latin_am, rest_world)

####RE-IMPORT####
comtrade_reim <- filter(comtrade, trade_flow==4)
reimports <- trade_flow_sort(comtrade_reim,core,output_regions,rest_asia,
   rest_europe, rest_latin_am, rest_world)

# ------------- End R File ---------------------------------------------
```

References

BC Government (2019) Fibre and mill information. Available at: https://www2.gov.bc.ca/gov/content/industry/forestry/competitive-forest-industry/forest-industry-economics/fibre-mill-information (accessed November 7, 2019).

Chen, X. and Önal, H. (2012) Modelling agricultural supply response using mathematical programming and crop mixes. *American Journal of Agricultural Economics* 94(3), 674–686.

CINTRAFOR (2014) Center for International Trade in Forest Products. University of Washington, School of Environmental and Forest Sciences, Seattle, Washington. Available at: https://www.cintrafor.org/data.html (accessed July 20, 2020).

de Frahan, B.H., Buysse, J., Polome, P., Fernagut, B., Harmignie, O. *et al.* (2007) Positive mathematical programming for agricultural and environmental policy analysis: review and practice. In: Weintraub, A., Romero, C., Bjorndal, T. and Epstein, R. (eds) *Handbook of Operations Research in Natural Resources.* Springer, New York.

EuroStat (2019) Eurostat. Your key to European statistics. Available at: http://epp.eurostat.ec.europa.eu/ (accessed October 16, 2019).

FAO (2020) FAO forestry statistics. Food and Agricultural Organization of the United Nations, Rome, Italy. Available at: http://www.fao.org/faostat/en/#data (accessed July 20, 2020).

Government of Canada (2019) National forestry database. Updated 16 April. Available at: http://nfdp.ccfm.org/index_e.php (accessed October 16, 2019).

Heckelei, T., Britz, W. and Zhang, Y. (2012) Positive mathematical programming approaches – recent developments in literature and applied modelling. *Bio-based and Applied Economics* 1(1), 109–124.

Howard, J.L. (2001) U.S. Timber Production, Trade, Consumption, and Price Statistics 1965 to 1999. *Research Paper FPL-RP-595*. April. U.S. Department of Agriculture, Forest Service, Forest Products Laboratory, Madison, Wisconsin.

Howitt, R.E. (1995) Positive mathematical programming, *American Journal of Agricultural Economics* 77, 329–342.

Johnston, C.M.T. and van Kooten, G.C. (2016) Global trade impacts of increasing Europe's bioenergy demand. *Journal of Forest Economics* 23, 27–44.

Lamers, P., Junginger, M., Hamelinck, C. and Faaij, A. (2012) Developments in international solid biofuel trade – an analysis of volumes, policies, and market factors. *Renewable and Sustainable Energy Reviews* 16, 3176–3199.

McCarl, B.A. (1982) Cropping activities in agricultural sector models: a methodological proposal. *American Journal of Agricultural Economics* 64(4), 768–772.

Niquidet, K. and Tang, J. (2013) Elasticity of demand for Canadian logs and lumber in China and Japan. *Canadian Journal of Forest Research* 43(12), 1196–1202.

Önal, H. and McCarl, B.A. (1991) Exact aggregation in mathematical programming sector models. *Canadian Journal of Agricultural Economics* 39(2), 319–334.

Oswalt, S.N. and Smith, W.B. (eds) (2014) *U.S. Forest Resource Facts and Historical Trends. USDA Forest Service FS-1035, August, Revised September*. U.S. Department of Agriculture, Forest Service, Washington, DC. Available at: https://www.fia.fs.fed.us/library/brochures/docs/2012/ForestFacts_1952-2012_English.pdf (accessed October 19, 2019).

Paris, Q. (1991) *An Economic Interpretation of Linear Programming*. Iowa State University Press, Ames, Iowa.

Paris, Q. (2011) *Economic Foundations of Symmetric Programming*. Cambridge University Press, Cambridge, UK.

Paris, Q., Drogué, S. and Anania, G. (2011) Calibrating spatial models of trade. *Economic Modelling* 28(6), 2509–2516.

Perez-Garcia, J.M. (1993) Global forestry impacts of reducing softwood supplies from North America. *CINTRAFOR Working Paper 43*. University of Washington, School of Environmental and Forest Sciences, Seattle, Washington.

Random Lengths (various years) *Forest Product Market Prices and Statistics Yearbook*. Published by Anderson, J.P., West, N., Fitzgerald, A. and Guzman, D. Random Lengths Publications Ltd, Eugene, Oregon.

UNECE (United Nations Economic Commission for Europe) (2019) Sustainable development goals. Available at: http://www.unece.org/fileadmin/DAM/ timber/docs/dp/dp-30.pdf (accessed October 16, 2019).

USDA Forest Service (2019) We are the nation's forest census. Available at: https://www.fia.fs.fed.us/ (accessed October 16, 2019).

van Kooten, G.C. and Johnston, C.M.T. (2014) Global impacts of Russian log export restrictions and the Canada–U.S. lumber dispute: modelling trade in logs and lumber. *Forest Policy and Economics* 39, 54–66.

Vercammen, J. (2011) *Agricultural Marketing. Structural Models for Price Analysis*. Routledge, London and New York.

WTO (2019) *Tariff download facility: WTO tariff database. World Trade Organization*. Available at: http://tariffdata.wto.org/ReportersAndProducts.aspx (accessed November 7, 2019).

5 Economic Analysis of a Softwood Lumber Quota Regime and a Policy to Subsidize Biomass Generation of Electricity

Fatemeh Mokhtarzadeh and G. Cornelis van Kooten

University of Victoria, Victoria, BC Canada

Abstract

The REPA spatial price equilibrium model developed in Chapter 4 is used to investigate the regional welfare impacts of a quota on exports of Canadian softwood lumber to the U.S. In the model, Canada is divided into seven regions and the U.S. into five regions, with the rest of the world constituting a 13th region; the model is calibrated to the bilateral trade flows that existed in 2016 when there was free trade in lumber. Various quota levels are examined in terms of their impact on producers and consumers in both countries. Canadian producers are found to be better off with a hard quota compared with free trade, although the quota leads to a reduction in market share while driving a wedge between Canadian and U.S. prices, both of which are aggravated with harder quotas. Overall, the loss of export sales to the U.S. is not recouped with sales to the rest of the world. The REPA model is also used to examine the impact of EU demand for wood pellets to generate electricity. Results indicate that pellet prices will approximately double.

In this chapter, we employ the REPA trade model described in Chapter 4 (this volume) to: (i) examine the impact of a voluntary export quota on Canadian sales of softwood lumber to the U.S.; and (ii) determine the effect of European subsidies on wood biomass used to generate electricity. U.S. lumber producers are primarily interested in reducing imports of softwood lumber from Canada. Export taxes or import duties are one means of reducing U.S. imports, but revenues accrue to either the respective Canadian or U.S. government. The welfare and income redistribution consequences of a quota can be examined using the multi-region but one product (lumber) version of the REPA trade model.

As an alternative, export quotas enable Canadian lumber producers to capture the rents associated with supply restrictions. European energy subsidies, on the other hand, will lead to an increase in the value of wood residuals. They will increase timber harvests and lumber production, thereby lowering lumber prices

10.1079/9781789248234.0005

but increasing the prices of products produced from wood residuals, as there is increased competition for the use of such residuals. To address that issue requires the use of the multi-region, multi-product version of the REPA trade model.

5.1 Modeling a Quota Regime

Van Kooten (2002) and Johnston and van Kooten (2017) discussed the inefficiency that exists in allocating quota across Canadian provinces. They argued that it is more efficient to allocate quota on the basis of a company's historic production or historic sales to the U.S. and then allow trade in quota. This will prevent situations where some provinces (such as Alberta in 2011) pay over-quota fees while others (BC interior and BC coast) are at the same time unable to reach their quota limits. Johnston and van Kooten's (2017) findings indicate that, in the absence of a softwood lumber agreement (SLA), Alberta would supply an additional 9% of softwood lumber to the U.S. Here, we are concerned about the overall level of a quota on Canadian softwood lumber and not how quota might be allocated among provinces. We assume that their share of exports to the U.S. determines the quota level for each province in 2016; we also assume that the provinces collect the quota rents (although the rent will accrue to the lumber producers in those jurisdictions).

Consider the quota model illustrated in Fig. 5.1. The U.S. demand function is the excess demand function – the difference between U.S. demand and U.S. supply for prices below the U.S. autarkic price. This is the effective demand for Canadian softwood lumber. The Canadian supply function is the difference between Canada's supply and demand for prices above the Canadian autarkic price (see van Kooten, 2002, 2020).

Under a quota regime, the excess supply curve becomes vertical at the quota amount q^Q. The quota raises the price received by U.S. lumber producers to P'_{US}, but it also increases what U.S. consumers pay. The net loss to U.S. consumers is measured by area (B+E). With a quota, the Canadian price falls below the free trade price P_C to P'_C. Hence, Canadian consumers are better off, apparently at the

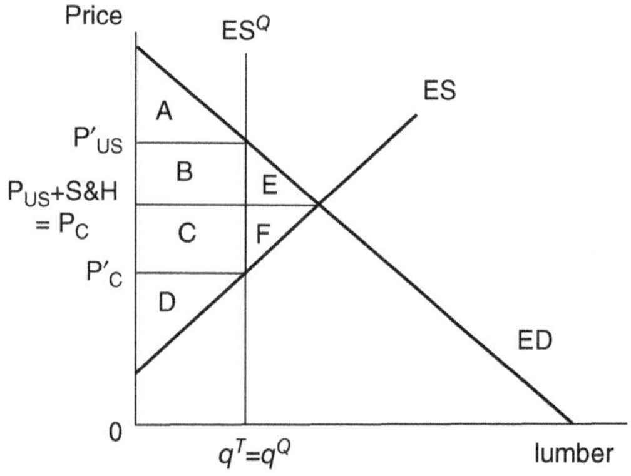

Fig. 5.1. International trade model of softwood lumber quota. (Authors' own figure.)

expense of domestic producers. The net loss to Canadian producers is given by area (C+F) resulting from the reduction in Canadian exports of softwood lumber to the U.S. However, there remains a price gap between P'_{US} and P'_{C} due to the trade restriction. This price wedge creates a quota rent or surplus given by area (B+C), which is captured by Canadian producers. In general, area B is quite a bit larger than area F, so Canadian producers benefit; the case is similar to that of a monopolist. Yet, this benefit could be partially dissipated by permitting higher-cost producers in one province to continue selling lumber to the U.S. while lower-cost producers in another province are forced to reduce or eliminate their output. This could be mitigated by allowing quota to be traded across provinces.

To quantify the impact of quota, we use the multi-region, single commodity version of the REPA spatial price equilibrium model. It allows us to determine the prices and quantities produced and consumed in each region, and bilateral flows of softwood lumber among regions. As in previous chapters, assume there are N exporting and M importing regions where q_{ij} is the quantity of trade between region i and j and τ_{ij} is the per unit shipping and handling (S&H) costs. To quantify the welfare effects, the objective is to maximize the total consumer and producer surpluses across regions plus any policy-induced scarcity rents by allocating lumber from excess supply regions/countries to excess demand regions/countries, while minimizing total S&H (transacation) costs. Hence, the objective function can be written to maximize total surplus (TS) as follows:

$$\text{Maximize}\left(q_{i,j}\right)\ \ \text{TS} = \sum_{j=1}^{M}\left(\alpha_j q_{ij} + 0.5\beta_j q_{ij}^2\right) - \sum_{j=1}^{N}\left(\mu_i q_{ij} + 0.5\delta_i q_{ij}^2\right) - \sum_{j=1}^{M}\sum_{j=1}^{N}\tau_{ij} q_{ij} \tag{5.1}$$

$$\text{Subject to:}\quad \sum_{j=1}^{M} q_{ij} \le q_i,\ \forall i \qquad [P_i^s] \tag{5.2}$$

$$q_j \le \sum_{i=1}^{N} q_{ij},\ \forall j \qquad [P_j^d] \tag{5.3}$$

A further requirement is that q_{ij}, q_i, $q_j \geq 0$, which is referred to as the non-negativity constraint. The shadow (dual) prices are provided in square brackets in constraints (5.2) and (5.3). The coefficients α_i, β_i, μ_j and δ_j are assumed to be known parameters that are determined from information on elasticities, production, consumption and prices in each region; demand and supply quantities are given, while q_{ij} are unknown and endogenously determined by the solver. It is also assumed that the S&H costs from exporting to importing countries are known. Constraints (5.2) state that an exporting region cannot supply more than it produces, while constraints (5.3) require that the demand of an importing region is satisfied. An alternative specification that minimizes costs is provided in Appendix 5.A.

5.1.1 Model data

The data employed in the model come from the UN's Food and Agricultural Organization (FAO), Statistics Canada, BC Stats, U.S. Forest Service and the publication *Random Lengths*. Significant contradictions were found in the data, but these were resolved by carefully comparing data from different sources and using averages where justified. Importantly, the data are based on 2016, because restrictions under the 2006 Softwood Lumber Agreement (SLA2006) did not come into effect that year and the market was characterized by free trade. In 2016, prices in the U.S. ranged from US\$137.92/m^3 (\$325.46/mbf) to \$182.84/m^3 (\$488.94/mbf). These are close to the 2016 Random Lengths' composite price for Eastern spruce-pine-fir (SPF) lumber (Northern U.S., \$370/mbf), the composite price for Southern pine (South, \$382/mbf), and an average of the Inland (\$406/mbf) and Western SPF (\$294/mbf) composite lumber prices for 2016. However, these composite prices are not used to construct Random Lengths' framing lumber composite price, which is the benchmark price used in the SLA. We divide Canada into seven regions (BC coast, BC interior, Alberta, Ontario, Quebec, Maritimes (Newfoundland and Labrador, New Brunswick, Prince Edward Island, and Nova Scotia) and Prairies (Saskatchewan and Manitoba) and the U.S. into five regions according to the U.S. Bureau of Economic Analysis (Northwest, Midwest, Southeast, Southwest, and West). Table 5.1 indicates how U.S. states are grouped into regions.

One of the key factors in the international market is the distance between trading partners, which determines S&H costs, among other things. (In Chapter 7, a gravity model structure is considered, because it explicitly models distances among trading regions.) Figure 5.2 indicates the allocation of softwood lumber exports from Canada and its regions to different regions in the U.S. Overall, about 48% of Canadian lumber ships to markets in the West and Midwest regions of the U.S. These two regions are the largest demand market for the BC coast (accounting for 58.1% of the BC coast's total exports), BC interior (58.1%), Alberta (54.5%), Prairies (83.1%), and Ontario (52.9%). The Northeast states of the U.S. are the destination of 52.0% of lumber exports from Quebec and 39.1% of exports from the Maritimes.

Table 5.1. U.S. regional classification (Authors' own table, adapted from: U.S. Bureau of Economic Analysis (BEA) at https://www.bea.gov/)

Region				
Midwest	Illinois Indiana Iowa	Kansas Michigan Minnesota	Missouri Nebraska North Dakota	Ohio South Dakota Wisconsin
Northeast	Connecticut Maine Massachusetts	New Hampshire New Jersey New York	Pennsylvania Rhode Island Vermont	
Southeast	Alabama Delaware Dist. Of Columbia	Florida Georgia Kentucky	Maryland Mississippi North Carolina	South Carolina Tennessee Virginia West Virginia
Southwest	Arkansas	Louisiana	Oklahoma	Texas
West	Alaska Arizona California	Colorado Hawaii Idaho	Montana Nevada New Mexico	Oregon Utah Washington, State Wyoming

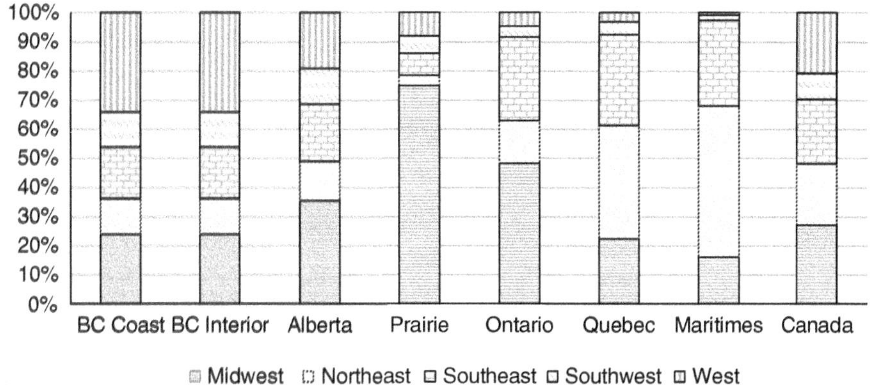

Fig. 5.2. Canadian export of softwood lumber to the U.S. by region, 2016. (Adapted from: Statistics Canada. Merchandise trade data, annual, and authors' calculations.)

Table 5.2 summarizes regional production, consumption, exports, imports, elasticities, and domestic prices. This provides the base-run bilateral trade flows, production, consumption, exports, and imports by region, all measured in cubic meters with prices in U.S. dollars.

In the analysis, we wish to determine the impact of ever tighter quotas and the extent and limits to which such tightening leads to greater revenues/rents for Canadian producers, assuming they can collect the quota rent. The quota rent, in turn, depends on the effect that the quota has on prices and quantities. Further, we compare the use of a quota instrument with an export tax and import tariff. Since prices and quantities are simultaneously determined in the model, a standard linear programming approach cannot be employed. Rather, it is necessary to employ linear mixed complementarity programming (MCP) (see Nicholson and Bishop, 2004).

The model was written and solved within an R-Excel-GAMS software environment. Results are extremely sensitive to the parameters employed in the model. The observed trade flows between regions are provided in Appendix 5.B (Table 5.B.1). The restructured REPA trade model that includes a hard quota component was then solved, with findings reported in the next section. The GAMS code is provided in Appendix 5.C.

5.1.2 Results

We consider a case where, rather than the United States imposing a TRQ on Canadian lumber, Canada agrees to a voluntary quota on sales of lumber to the U.S. Economic theory is clear that, if Canadian exports have an impact on prices in the U.S. (i.e. Canada is able to exercise market power), then Canadian lumber producers can maximize their net earnings by reducing their overall output. That is, by reducing output Canada can take advantage of its monopoly

Table 5.2. Softwood lumber production, export, prices and elasticity of supply, 2016. (Authors' own table.)

Regions	Production ('000s m³)	Exports ('000s m³)	Consumption ('000s m³)	Imports ('000s m³)	Price (US$/m³)	Price elasticity of Supply	Price elasticity of Demand
BC Coast	3,771.8	3,533.7	2,528.5	2,290.4	155.6	0.84	–0.2
BC Interior	27,585.5	27,161.8	555.0	131.3	135.0	0.38	–0.5
Alberta	9,281.0	6,571.0	2,718.8	8.8	133.3	0.38	–0.2
Ontario	3,455.0	2,333.5	8,985.3	7,863.8	122.3	0.65	–0.4
Quebec	13,817.7	8,547.9	5,413.8	144.0	133.5	0.65	–0.6
Maritimes	5,919.4	4,427.9	1,545.1	53.6	145.6	0.65	–0.8
Prairie	768.0	580.1	994.7	806.7	142.8	0.45	–0.8
Midwest	2,302.3	1,847.3	20,231.4	19,776.5	182.8	0.31	–0.3
Northeast	5,881.5	4,612.5	18,300.6	17,031.6	137.9	0.31	–0.8
Southeast	16,684.5	14,485.7	14,506.5	12,307.6	176.5	0.31	–0.3
Southwest	6,852.8	6,097.1	9,754.1	8,998.3	207.2	0.31	–0.3
West	19,201.6	14,338.9	21,079.6	16,216.9	170.3	0.31	–0.3
ROW[a]	334,086.6	4245.8	342,994.2	13,153.4	198.0	0.70	–0.17

[a] Countries outside the U.S. and Canada are aggregated into ROW (Rest of the World).

position as a major softwood lumber supplier to the U.S. market. Canada's domestic lumber producers can benefit greatly from a quota if they are able to collect the entire surplus resulting from the quota (areas B and C in Fig. 5.1). The sweet spot where producers maximize their net income is determined by the point where the marginal benefit associated with the U.S. excess demand function equals the marginal cost of Canadian lumber producers. Any output below or above this amount will lead to lower net returns. To examine this phenomenon in the case of a hard quota, we use Canada's share of U.S. consumption as a benchmark for computing the quota of each region in Canada. We examine hard quotas of 39%, 34%, 30% and 28% of U.S. consumption, and fix provincial quotas on the basis of their share of exports to the U.S. market in 2016. The U.S. total demand in 2016 was 84,847,210 m^3 (see Table 5.B.1) if Canadian lumber producers supplied 39%, the overall quota available to Canadian producers would be 33,090,411.71 m^3. If this is subsequently allocated across regions, it would be done as follows: BC Coast, 6.2%; BC Interior, 45.7%; Alberta, 8.3%; Ontario, 8.5%; Prairies, 1.7%; Quebec, 20.1%. No quota is imposed on producers in the Maritimes.

We find that with a 39% hard quota, 47.6% of Canada's domestic production is exported to the U.S., which is 5.8% lower than Canada's share under a free market (53.4%). Canada's share of U.S. consumption declines as the quota on Canadian provinces is tightened, with the lowest market share at 35.5% with a 28% hard quota. This confirms the results of van Kooten (2002) where a hard quota reduces Canada's exports to the U.S. but leads to significant gains to Canada. Of course, a more restrictive quota would benefit U.S. producers even more, as the quota raises U.S. producers' production and their share of the domestic market. These results are evident from Table 5.3. The U.S. producers' share of domestic consumption increases from 69.8% under a free market to higher levels as the quota is tightened. The production to consumption ratio peaks at 73.4% under a 28% hard quota.

A quota creates a wedge between Canadian and U.S. prices that exceeds the S&H costs. The price gap in the free market is on average \$34.92/m^3, which represents the S&H cost. With a hard quota of 39%, the price gap increases by 17.8% (\$41.12/m^3), thereafter by 28.4% (\$44.83/m^3), 34.5% (\$46.98/m^3) and 41.5%

Table 5.3. Estimate of export to production ratio, production to consumption ratio, and price gap[a]. (Authors' own table.)

Quota scenario	Canada export to production ratio[a] (%)	U.S. production to consumption ratio (%)	Price gap[b] (\$/m^3)
Free trade	53.39	69.81	34.92
39%	47.58	69.44	41.12
34%	41.82	71.39	44.83
30%	37.05	72.80	46.98
28%	35.51	73.36	49.42

[a] This is the ratio of total Canadian exports from all regions to total production of softwood lumber.
[b] Measured as the difference between U.S. consumer prices and Canadian producer prices, where prices are averaged across U.S. and Canadian regions, respectively.

($49.42/m^3), respectively, with quotas of 34%, 30%, and 28%. The benefits of the price gap are captured by Canadian producers in the form of quota rent. Theory predicts that a tighter quota policy leads to higher quota rents for Canadian producers, which is the case here. Indeed, as long as the gain in quota rent – area (B+C) in Fig. 5.1 – exceeds the loss in quasi-rent – area (C+F) in Fig. 5.1 – Canadian producers benefit from a voluntary quota. As the quota is tightened, the difference given by area B minus area F will first increase and eventually decline. Canada could determine at what point the quota rent reaches a maximum and thereby extract the greatest benefit from acting as a monopoly seller in the U.S. lumber market. The quota rents for different quota levels are provided in Table 5.4. The results indicate that, by reducing the share of Canadian lumber in the U.S. market from 39% to 28%, Canadian producers can gain more than US$225 million.

With a 39% quota, each Canadian region seeks to maximize its income by exporting as much as possible to each of the U.S. regions to which its quota applies. British Columbia's lumber producers earn a quota rent of $251 million over and above what they would have earned if they operated in a free market, followed by Quebec ($47 million), Ontario ($11 million) and Alberta ($6 million). When the quota is restricted so Canada accounts for less than 39% of the U.S. market, the price gap widens (Table 5.3) and quota rents increase (Table 5.4). The quota rent that accrues to BC increases by 2.6% ($245 million), 26.7% ($318 million) and 39.9% ($351 million) at quota levels of 34%, 30%, and 28%, respectively. Other provinces will likewise benefit, except perhaps provinces in Atlantic Canada.

Compared with free trade, projected changes in the wellbeing of U.S. consumers and Canadian producers, along with changes in other key important variables, are provided in Table 5.5. With a hard quota that limits Canada to 39% of the U.S. market for softwood lumber, Canadian shipments to the U.S. fall by 1.1 million m^3, or 3.3%, leading to a reduction in the Canadian lumber price of 4.3%. Overall, Canada's producers experience a reduction in quasi-rent of 7.3%, although this is more than offset by the gain in quota rent. With stricter quota levels, Canadian exports of lumber fall continuously (a reduction by 19.5% or $7.9 million under a 28% hard quota). Much of the Canadian lumber is simply replaced by lumber from U.S. suppliers, although overall consumption declines

Table 5.4. Quota rents for different hard quota levels on U.S. imports of Canadian softwood lumber, as proportion of U.S. consumption (US$ '000s). (Authors' own table.)

	Hard quota level			
Regions	39%	34%	30%	28%
BC Coast	482	2,661	12,369	18,114
BC Interior	250,896	242,370	305,813	333,600
Alberta	6,463	9,994	25,789	32,763
Ontario	11,162	11,699	23,597	28,178
Quebec	47,927	57,493	112,389	126,262
Prairies	310	1,065	4,015	5,697
Total Canada	317,239	325,282	483,971	544,615

from about 109 million m^3 under free trade to 107 million m^3 under a 28% hard quota, or by 1.9 million m^3 – a reduction of about 1.7%.

Ontario's production in 2016 was 3.45 million m^3 of which 2.3 million m^3 was exported to the U.S. Its consumption of 8.9 million m^3 is the highest among Canadian regions, with most imported from other regions in Canada (about 5.7 million m^3). U.S. trade measures restrict Ontario's ability to import and export compared with free trade. Ontario's exports to the U.S. are projected to decline to below 1.5 million m^3 at various quota levels, while other provinces find better opportunities to sell lumber to the U.S. Nonetheless, Ontario's production is projected to increase by about 4.5 million m^3 in response to increased domestic demand. Therefore, we excluded Ontario production in calculating the values in Table 5.5, as well as production by the Atlantic Provinces.

Meanwhile, Canadian lumber production from the affected provinces is projected to fall from 55 million m^3 to 54 million m^3, or by 763,000 m^3 (1.4%) with a 28% hard quota. However, with a 39% quota, we find a slight increase in lumber production caused by an increase in output from the BC Interior of 0.4%. Canada's lumber prices decline steadily as the hard quota becomes more restrictive compared with free trade – from 4.1% under a 39% hard quota to 5.7% under a 28% hard quota. The loss in producer surplus peaks at 24.4% (reduced by $1.24 billion) with a 28% hard quota; but, as mentioned above, this is offset by the gain in quota rent.

As a whole, the economic welfare of consumers in the U.S. declines due to higher prices. It is apparent from the results that prices in the U.S. increase continuously as the quota on Canadian sales to the U.S. increase; prices increase by $0.16/m^3 (0.1%) with a 39% quota and by $6.51/m^3 with a 28% quota. Accordingly, consumer demand falls, leading to a huge decline in consumer surplus by 27.9% with a 28% hard quota in comparison with free trade.

Table 5.5. Estimated changes in prices, consumption, production, exports, Canadian producer surplus, and U.S. consumer surplus relative to free trade[a]. (Authors' own table.)

	Canada			
Hard Quota	Price	Production	Export to U.S.	Producer Surplus
39%	–4.31%	0.15%	–3.33%	–7.28%
34%	–4.44%	–0.68%	–12.96%	–15.90%
30%	–5.39%	–1.12%	–19.24%	–24.36%
28%	–5.71%	–1.38%	–19.45%	–24.40%
	United States			
	Price	Consumption	Production	Consumer Surplus
39%	0.09%	0.98%	0.46%	–5.28%
34%	2.11%	–1.74%	0.50%	–21.28%
30%	2.58%	–2.40%	1.78%	–22.82%
28%	3.72%	–1.72%	3.29%	–27.91%

[a] Prices are measured as the average price across regions. Production and consumption measured as the sum across regions.

We find that Canadian producers are better off with a quota than under free trade, with the quota rent more than offsetting any loss in quasi-rent. Further, the wedge between Canadian and U.S. prices increases as the restriction on Canadian exports becomes stronger, leading to losses in U.S. consumer surplus to the benefit of U.S. producers. Overall, a hard quota on softwood lumber sales to the U.S. benefits Canadian producers if they capture the policy-induced scarcity (quota) rent, Canadian consumers (as the price in Canada is lowered), and U.S. lumber producers, who receive a higher price for their output. The big loser is the U.S. consumer.

5.2 Climate Change and Energy Policy: European Demand for Wood Biomass

At the 21st Conference of the Parties (COP-21) to the 1992 UN Framework Convention on Climate Change (FCCC) held in Paris during December 2015, 195 countries agreed to reduce carbon dioxide (CO_2) plus CO_2-equivalent greenhouse gas (GHG) emissions (hereafter simply CO_2 emissions) so that the projected increase in average global temperature would be kept below 2°C. Much like the 1997 Kyoto Protocol, the Paris Agreement relies on countries to meet their self-declared targets, known as Nationally Determined Contributions (NDCs), with shaming as the main mechanism to enforce compliance. However, NDC targets are sufficiently vague so that many countries can meet their intended obligations *de jure*, while avoiding them in practice – *de facto*. NDCs for each country are found at http://www4.unfccc.int/submissions/indc/Submission%20 Pages/submissions.aspx (accessed June 18, 2020).

Consider some examples. Russia agreed to reduce CO_2 emissions by 70–75% from their 1990 level by 2030, but actual emissions will be reduced by only 20–30% (a target already met) with the remainder to come from rational forest use, protection, maintenance, and regeneration. The U.S. would "make best efforts" to reduce emissions by 26–28% from a 2005 baseline by 2025. China agreed to *begin* reducing CO_2 emissions no later than 2030; to lower CO_2 emission intensity (emissions per unit of GDP) by 60% to 65% from the 2005 level by 2030; to increase the share of non-fossil fuels in primary energy consumption to around 20%; and to sequester an additional 900 gigatonnes (Gt) of CO_2 in forests between 2005 and 2030.

Canada committed to reduce its emissions by 30% below 2005 levels by 2030. It intends to do this mainly by preventing construction of new coal-fired power plants, phasing out existing coal plants or converting them to co-fire with biomass, implementing regulations on vehicle fuel standards, and relying on forestry activities. To avoid too great a disruption of the economy, Canada will place greater emphasis on forestry to sequester carbon from the atmosphere, substitute wood for steel/concrete in construction, and substitute residual biomass for fossil fuels as an energy source.

The European Union (EU) committed to "a binding target of at least 40% domestic reduction in greenhouse gas emissions by 2030 compared to 1990." This target is "in line with the EU objective ... to reduce its emissions by 80–95% by 2050 compared to 1990," pointing out in its INDC submission that this latter target had been agreed to by the developed countries (see European Commission, 2014). This was at the G8 meeting in L'Aquila, Italy in 2009 where countries agreed to limit the increase in global average temperature to 2°C by reducing global CO_2 emissions by 50% and their own emissions by 80% or more by 2050. More recently, the EU has made it clear that its ultimate objective is to eliminate use of fossil fuels by 2050.

Forest biomass is expected to be the most significant future source of renewable energy within the EU, and likely to account for over half of total renewable energy production (European Commission, 2013). Yet little is known about how increased demand for biomass in the EU will impact the global forest products sector. Timber-rich regions will undoubtedly benefit from increased demand for wood fiber, but other wood product markets may experience significant changes in prices from increased competition for residual and other fiber. Early research indicates that residual fiber prices will more than double, although lumber prices are likely to fall as sawmill residues become more valuable (Johnston and van Kooten, 2015, 2016); a reduction in lumber prices will, in turn, incentivize substitution of lumber for concrete and steel in construction, thereby lowering CO_2 emissions (van Kooten and Johnston, 2016; van Kooten, 2018).

Johnston and van Kooten (2016) modified the multi-region, multi-product REPA forest trade model to examine the impact that European subsidies on wood pellets for use in power plants would have on global forest product markets. The model was described in Chapter 4 and was based on Fig. 3.6 in Chapter 3. It consists of 20 regions (including five Canadian regions and three U.S. regions) and eight final products. Using the REPA model, Johnston and van Kooten (2016) investigated the effects that a doubling of European demand for wood pellets to generate electricity would have on global prices of a variety of wood products. Depending on the region, prices of the following wood products were found to increase: industrial roundwood by 1%, particleboard by 7%–16%, fiberboard by 5%–9%, wood pulp by 8%–12%, and wood pellets by 71%–128%. Meanwhile, prices of sawnwood and plywood and veneer were projected to fall by 1% and 4%–7%, respectively, because more logs would be harvested.

Timber-producing regions were projected to gain from the European policy, and particularly Russia, Canada, and the U.S. would likely experience large net welfare gains, while Asia would be a net loser. As expected, consumers of electricity and/or taxpayers in the EU and other regions implementing renewable energy policies are the biggest losers. As a result of the European policy, the price of wood pellets is projected to nearly double, rising to between \$107 and \$154 per tonne. Overall, however, the model demonstrates the need to account for the global interconnections among forest products consumers and producers.

5.3 Concluding Discussion

To address issues related to softwood lumber trade, it is useful to employ economic methods appropriate to the problem at hand. Methods need to be rooted solidly in economic theory. The REPA global trade model is a spatial price equilibrium model that is suitable for the types of applications addressed in this study (see also Johnston and van Kooten, 2014). The current version of the model employs linear mixed complementarity programming, which is required for solving regional equilibrium models that investigate the impacts of trade policies.

The effect of the different quotas considered in this chapter is to reduce Canadian exports of softwood lumber to the U.S. to below what would occur under free trade. This raises the price received by U.S. lumber producers at the expense of U.S. consumers. However, depending on their level, quotas create a form of monopoly rent, which we refer to as a policy-induced (quota) rent. Since U.S. lumber producers appear unconcerned about this rent and are unable to capture it, the dispute from the Canadian side really concerns the potential size of the rent and how it is allocated or dissipated. These outcomes accord with economic theory. The analysis in this chapter suggests that a quota system is generally beneficial to producers in Canada, since the quota rent more than compensates for the loss in quasi-rent (also known as producer surplus). Lumber producers in various Canadian provinces should prefer a quota to an export tax and/or a TRQ approach (van Kooten, 2002, 2014; Johnston and van Kooten, 2017).

The quota is easy to administer and enables lumber producers to collect rents as opposed to government. However, the quota approach can be abused in the absence of trade in quota, because it permits higher-cost producers to remain in the industry where a free market would have driven them to compete. This is exacerbated across provinces because low-cost producers in one province may not have sufficient quota to achieve economies of scale, while high-cost producers in another province continue to sell to the U.S. Therefore, it is imperative in any quota system to permit quota trade that will compensate high-cost Canadian producers while ensuring that lumber for export to the U.S. is produced at lowest possible cost. Overall, the results provided in this study are representative of what is likely to happen if Canada imposes a voluntary quota regime on lumber exports to the U.S. Based on our analysis and on economic theory, Canada should welcome quotas while seeking to displace U.S. lumber in overseas markets.

Finally, a multi-region, multi-product version of the REPA trade model (see Chapter 4) was used to examine the global impacts of European subsidies on biomass-burning to generate electricity. Since the energy use of wood biomass competes with pulp, medium-density fiberboard (MDF), oriented strand board (OSB), and other products that use sawmill and logging residuals, EU subsidies for wood pellets lead to important consequences across all regions. First, the subsidies raise the price of residual fiber, resulting in an increase in residual prices that offset some of the benefits of the subsidy while increasing the price of other commodities that use such fiber. Second, the value of raw logs increases so that more timber is harvested. Further, since most residuals are associated with sawmilling as opposed to logging (as it is costly to gather and transport logging residues that are also environmentally important to forest ecosystem productivity and diversity), more lumber is produced. This causes the price of lumber to decline. Third, there is a transfer of wealth to countries that are considered timber rich. Whether these developments are to be welcomed is a political decision, as are most other forest conservation and exploitation policies.

APPENDIX 5.A: Alternative Model Specification

Suppose q_{ij} represents the quantity exported from region i to region j, and τ_{ij} is the associated per unit S&H costs. An alternative objective for a region in the international trade market is to minimize the total transportation cost rather than overall welfare (sum of consumer plus producer surpluses minus transportation costs). The objective function and constraints (along with the dual variables) are specified as follows:

$$\min_{q_{ij}} c = \sum_{i=1}^{I} \sum_{j=1}^{J} t_{ij} q_{ij}$$

Subject to: Dual Variable

(1) $\sum_{j=1}^{J} q_{ij} \leq (\mu_i + P_i) / \delta_i$ $[P_i]$

(2) $(\alpha_j - P_j) / \beta_j \leq \sum_{i=1}^{I} q_{ij}$ $[P_j]$

(3) $q_{ij}, q_i, q_j \geq 0$ (non-negativity)

The coefficients α_i, β_i, μ_j and δ_j are parameters that are calculated based on elasticities of supply and demand, and information on domestic supply and demand quantities. Available data on bilateral trade can be used to calibrate the model, but q_{ij} are unknown and must be determined endogenously by the SPE model for different policies. It is also assumed that S&H costs from exporting to importing countries are known. The first constraint states that an exporting region cannot supply more than its production and the second constraint states that the demand of an importing region must be satisfied. The dual variables in this case are the prices in the exporting and importing regions.

APPENDIX 5.B: Expanded Results

Table 5.B.1. Observed softwood lumber trade flows among net exporting and net importing countries ('000s m^3), 2016[a]

Export/Import Regions	BC Coast	BC Interior	Alta	Ontario	Quebec	Maritime	Prairie	Mid-west	North-east	South-east	South-west	West	Rest of World
BC Coast	238.1	0	0	0	0	0	0	517	265.5	380.9	256.7	734.8	1378.9
BC Interior	1692.2	423.7	0	1312.2	0	0	0	3780.3	1941.2	2785	1877	5373.4	8400.4
Alberta	0	0	2710	1137.3	0	0	794.3	1535.5	579.9	859.9	527.7	820.2	316.2
Ontario	0	0	0	1121.5	0	0	0	1137.9	307.7	644.5	103.3	127.6	12.5
Quebec	0	0	0	1772.2	5269.8	0	0	1501	2616.3	2078.5	293.6	211.1	75.3
Maritimes	0	0	0	1410.7	0	1491.5	0	483.4	1560.6	879.6	55	23.6	14.9
Prairie	0	0	0	27	0	0	187.9	413.8	18.9	41.8	32.2	44.1	2.1
Midwest	0	0	0	237.1	0	0	12.4	455	461.2	276.8	276.8	575.7	7.4
Northeast	0	0	0	0	0	53.6	0	1251.9	1269	761.7	761.7	1584.1	199.5
Southeast	0	0	0	0	0	0	0	3614.2	3663.5	2198.9	2198.9	4573.2	435.9
Southwest	0	0	0	0	0	0	0	1242.2	1259.1	755.7	755.7	1571.8	1268.3
West	598.2	131.3	8.8	0	144	0	0	3843	3895.4	2338.1	2338.1	4862.7	1042
Rest of World	0	0	0	1967.2	0	0	0	456.2	462.4	505.1	277.6	577.3	329840.8

[a] Calculated by the authors using data from Comtrade (2019), FAO (2020), U.S. International Trade Commission (2017, 2019), BC Government (2019), Government of Canada (2019), Johnston and van Kooten (2017), and Johnston and Parajuli (2017).

Table 5.B.2. Total bilateral shipments of softwood, including Canada to the U.S., by region, free trade ('000s m^3). (Authors' calculations.)

Export/Import Regions	BC Coast	BC Interior	Alberta	Ontario	Quebec	Maritime	Prairie	Mid-west	North-east	South-east	South-west	West	Rest of World
BC Coast	0	20.64	22.33	33.38	22.18	10.07	12.88	27.20	17.72	20.84	51.56	14.70	42.36
BC Interior	20.64	0.00	1.69	12.74	1.54	10.57	7.76	47.84	2.92	41.48	72.20	35.34	63.00
Alberta	22.33	1.69	0.00	11.05	0.15	12.26	9.45	49.53	4.61	43.17	73.89	37.03	64.69
Ontario	33.38	12.74	11.05	0.00	11.20	23.32	20.50	60.58	15.66	54.22	84.94	48.08	75.74
Quebec	22.18	1.54	0.15	11.20	0.00	12.12	9.31	49.38	4.46	43.02	73.74	36.88	64.54
Maritimes	10.07	10.57	12.26	23.32	12.12	0.00	2.81	37.27	7.65	30.91	61.63	24.77	52.43
Prairie	12.88	7.76	9.45	20.50	9.31	2.81	0.00	40.08	4.84	33.72	64.44	27.58	55.24
Midwest	27.20	47.84	49.53	60.58	49.38	37.27	40.08	0.00	44.92	6.36	24.36	12.50	15.16
Northeast	17.72	2.92	4.61	15.66	4.46	7.65	4.84	44.92	0.00	38.56	69.28	32.42	60.08
Southeast	20.84	41.48	43.17	54.22	43.02	30.91	33.72	6.36	38.56	0.00	30.72	6.14	21.52
Southwest	51.56	72.20	73.89	84.94	73.74	61.63	64.44	24.36	69.28	30.72	0.00	36.86	9.20
West	14.70	35.34	37.03	48.08	36.88	24.77	27.58	12.50	32.42	6.14	36.86	0.00	27.66
ROW	42.36	63.00	64.69	75.74	64.54	52.43	55.24	15.16	60.08	21.52	9.20	27.66	0.00

Table 5.B.3. Canadian regions' exports to the U.S. under hard quota (m³). (Authors' calculations.)

Region	Free trade	Hard quota			
		28%	30%	34%	39%
BC Coast	2,154,785	1,462,675	1,567,152	1,557,293	1,786,307
BC Interior	15,756,978	10,726,208	11,492,366	13,024,681	14,940,076
Alberta	4,323,174	1,712,297	1,834,604	2,079,218	2,384,985
Ontario	2,320,916	1,236,158	1,324,455	1,303,275	1,355,336
Quebec	6,700,451	4,799,594	4,917,114	5,828,079	6,685,149
Maritimes	3,002,245	6,634,332	6,556,666	6,056,979	5,982,287
Prairies	550,906	393,124	421,204	449,455	515,551
Total Canada	**34,809,455**	**26,964,389**	**28,113,561**	**30,298,981**	**33,649,691**

Table 5.B.4. Quota rents when quota on U.S. imports of Canadian softwood lumber fixed at 39% of U.S. consumption (US$). (Authors' calculations.)

Canadian region	Midwest	Northeast	Southeast	Southwest	West	Total
BC Coast	152,492		112,342		216,754	481,588
BC Interior	88,648,137	10,097,667	823,836	64,245,183	87,081,371	250,896,194
Alberta	20,862	6,418,973	11,683		11,143	6,462,660
Ontario	11,161,938					11,161,938
Quebec	467,229	46,746,827	646,966		65,703	47,926,726
Prairies	128,332	154,823	12,974		13,662	309,791
Maritimes						0
Total Canada						**317,238,897**

Table 5.B.5. Quota rents when quota on U.S. imports of Canadian softwood lumber fixed at 34% of U.S. consumption (US$). (Authors' calculations.)

Canadian region	Midwest	Northeast	Southeast	Southwest	West	Total
BC Coast	842,524		620,696		1,197,579	2,660,799
BC Interior	82,486,558	16,009,774	4,551,735	56,008,621	83,313,541	242,370,229
Alberta	1,418,934	7,022,763	794,621		757,924	9,994,242
Ontario	11,698,523					11,698,523
Quebec	2,581,467	50,974,227	3,574,522		363,014	57,493,230
Prairies	709,039	208,717	71,684		75,484	1,064,924
Maritimes						0
Total Canada						325,281,946

Table 5.B.6. Quota rents when quota on U.S. imports of Canadian softwood lumber fixed at 30% of U.S. consumption (US$). (Authors' calculations.)

Canadian region	Midwest	Northeast	Southeast	Southwest	West	Total
BC Coast	3,481,807	1,373,288	2,565,082		4,949,107	12,369,283
BC Interior	92,863,737	42,663,227	18,810,461	49,419,372	102,056,145	305,812,942
Alberta	6,657,731	11,846,133	3,728,412		3,556,228	25,788,504
Ontario	17,915,551	5,681,633				23,597,183
Quebec	10,668,148	85,448,648	14,772,038		1,500,191	112,389,025
Prairies	2,930,170	476,171	296,239		311,943	4,014,524
Maritimes						0
Total Canada						**483,971,461**

Table 5.B.7. Quota rents when quota on U.S. imports of Canadian softwood lumber fixed at 28% of U.S. consumption (US$). (Authors' calculations.)

Canadian region	Midwest	Northeast	Southeast	Southwest	West	Total
BC Coast	5,119,893	1,944,739	3,771,876		7,277,513	18,114,020
BC Interior	100,387,567	44,680,999	27,660,219	46,124,747	114,746,803	333,600,334
Alberta	9,905,752	12,018,938	5,547,344		5,291,158	32,763,193
Ontario	21,907,080	6,270,866				28,177,945
Quebec	15,687,192	86,647,388	21,721,839		2,205,985	126,262,406
Prairies	4,308,727	494,178	435,611		458,703	5,697,219
Maritimes						0
Total Canada						544,615,117

Table 5.B.8. Estimated trade flow at 39% hard quota.

Region	bcc	bci	alt	ont	que	mrt	pra	mw	ne	se	sw	we	row	Production	Export
bcc	80,682							488,779	1,840,585	360,088	242,681	694,760	1,898,460	3,765,449	3,684,767
bci	2,448,828	555,677				1,547,731	708,727	3,584,352	364,388	2,640,627	1,779,645	5,094,866	7,495,118	27,696,156	27,140,479
alt			2,718,902	4,176,541				964,871		540,340		515,386		9,280,428	6,561,526
ont				2,505,345				1,355,336	2,610,336					3,860,681	1,355,336
que				1,654,201	5,421,348		36,006	1,497,603	5,982,287	2,073,710	292,902	210,598		13,796,704	8,375,356
mrt									18,834					5,982,287	5,982,287
pra							251,694	411,340		41,586		43,791		767,245	515,55,
mw								3,406,200	9,27,1,988					3,406,200	–
ne														9,271,988	–
se								12,539,092		12,915,100				25,454,192	12,539,092
sw											9,862,257			9,862,257	–
we										5,572,897		22,953,514		28,526,410	5,572,897
row											25,543		333,692,522	333,718,064	25,543
Consumption	2,529,510	555,677	2,718,902	8,336,087	5,421,348	1,547,731	996,427	24,247,572	20,088,419	24,144,348	12,203,027	29,512,914	343,086,099	475,388,,61	71,752,835
Import	2,448,828	–	–	5,830,742	–	1,547,731	744,733	20,841,372	10,816,431	11,229,249	2,340,770	6,559,400	9,393,578	71,752,835	
Price	150	129	128	139	127	139	137	177	154	171	201	165	198		

Table 5.B.9. Estimated trade flow at 34% hard quota.

Region	bcc	bci	alt	ont	que	mrt	pra	mw	ne	se	sw	we	row	Output	Export
bcc	238098							426115		313923	211568	605688	1936159	3731550	3493563
bci	2228478	559100				1555490	691950	3124820	1604643	2302085	1551485	4441678	9372273	27431973	26872872
alt			2725694	4431463				841170	317672	471065		449311		9236374	6510680
ont				2526818				1181575	121700					3830093	1303275
que	68345	0		1426761	5461878	6381		1305603	2275678	1807850	255350	183598	893192	13684636	8222758
mrt									6056979					6056979	6056979
pra							313759	358604	16419	36255		38177		763214	449455
mw								3406200						3406200	0
ne									9328642					9328642	0
se								6241022		19213170				25454192	6241022
sw											9837675			9837675	0
we								7362465				21163945		28526410	7362465
row											376359		331374871	331751229	376358
Consumption	2534921	559100	2725694	8385042	5461878	1561870	1005709	24247572	199721704	24144348	12232462	26882397	343576495	473039166	66889318
Import	2296823	0	0	5858224	0	1561870	1561870	20841372	10393062	4931179	2394761	5718451	12201624	66889318	
Price	154	133	132	143	131	144	141	183	158	176	205	170	196		

Table 5.B.10. Estimated trade flow at 30% hard quota.

Region	bcc	bci	alt	ont	que	mrt	pra	mw	ne	se	sw	we	row	Production	Export
bcc	760037							375984	193070	276991	186677	534430	1378918	3706107	2946071
bci	1778945	561670				1572483	673692	2757194	1415835	2031252	1368958	3919128	11255767	27334923	26773253
alt			2592323	4776382				742209	280299	415646		396451		9203309	6610986
ont				2482679				1042566	281889					3807134	1324455
que			138468	1162724	5492299			1152002	2007951	1595162		161998	1889916	13600522	8108223
mrt									6556666					6556666	6556666
pra							338984	316415	14488	31989	24627	33685		760188	421204
mw								3441043						3441043	0
ne									9707657					9707657	0
se								6178283		19545677				25723959	6178283
sw											9819224			9819224	0
we								8001837				20837799		28839636	8001837
row											855024		329419967	330274991	855024
Consume	2538982	561670	2730792	8421785	5492299	1572483	1012675	24007533	20457854	23896717	12254509	25883491	343944569	472775358	67776000
Imports	1778945	0	138468	5939106	0	1572483	673692	20566489	10750197	4351040	2435286	5045692	14524602	67776000	
Price	148	127	126	137	126	138	135	175	146	169	200	163	197		

Table 5.B.11. Estimated trade flow at 28% hard quota.

Region	bcc	bci	alt	ont	que	mrt	pra	mw	ne	se	sw	we	row	Production	Export
bcc	849433							350918	180198	258525	174232	498802	1378918	3691027	2841594
bci	1222712	239402				1254982		2573381	1321446	1895835	1277694	3657852	13834097	27277401	27038000
alt			2733813	4737601				692728	261612	387936		370021		9183711	6449898
ont				2557369				973062	263096					3793527	1236158
que	103973	323791		1148593	5510329	323791	1016805	1075202	1874088	1488818	210288	151199	323791	13550667	8040338
mrt									6634332					6634332	6634332
pra	365271							295321	13522	29857	22985	31440		758395	758395
mw								3467543						3467543	0
ne									9766567					9766567	0
se								6690026		19239103				25929129	6690026
sw											9808288			9808288	0
we								7706792		408310		20962755		29077856	8115102
row											774105		328625920	329400025	774105
Consume	2541389	563193	2733813	8443563	5510329	1578773	1016805	23824973	20314861	23708383	12267592	25672068	344162726	472338467	68577945
Imports	1691955	323791	0	5886194	0	1578773	1016805	20357429	10548294	4469280	2459304	4709313	15536806	68577945	
Price	152	131	130	141	129	142	139	193	180	187	203	181	194		

APPENDIX 5.C: GAMS Code for Modelling Quota Regime

```
$TITLE Softwood Lumber Trade Model-2016 Free Trade Model
* Model is solved in MCP Framework

$OFFSYMLIST OFFSYMXREF
OPTION LIMROW = 0
OPTION LIMCOL = 0
OPTION RESLIM = 50000
OPTION ITERLIM = 1000000
* OPTION MCP = NLPEC
$oneolcom
$eolcom #

SETS
  s supply regions ;
alias(s,d);
* -------------------------------------------------------
* SCENARIO TRIGGERS: DEFAULT IS FREE TRADE and REGIONAL QUOTA
* -------------------------------------------------------
SCALARS
  Qpercent Percent of US demand allocated to Canada in 2016 /0.34/ # 0.24
  0.28 0.30 0.34 0.39
  UScon      US consumption for determining quota /83806725/
  exchange   Canadian dollar as proportion of US dollar 2016 /0.754797/
  conv       Conversion factor m3 /1/ # Otherwise 2.46 m3 per mmbf
;
* ----------------------------------------------------------
* IMPORT DATA FROM TWO SPREADSHEETS; 2ND ONE IS CALIBRATED
DATA
* ----------------------------------------------------------
* Build gdxxrw instructions file
$onecho > import.txt
par=data rng=ModelData!B28:I41 Rdim=1 Cdim=1
par=ship rng=matrix!A1:N14 Rdim=1 Cdim=1
par=lumtax rng=matrix!A20:N33 Rdim=1 Cdim=1
par=lumflow rng=ModelData!B5:O18 Rdim=1 Cdim=1
set=s rng=ModelData!B6:B18 Rdim=1 Cdim=0
$offecho

* Create and import GDX file with import data
$call gdxxrw.exe TradeData-New.xlsx o=ImportData.gdx @import.txt
$gdxin ImportData.gdx

PARAMETERS
  data(s,*)    Supply data
  ship(s,d)    Bilateral shipping and handling costs $ per cubic meter
```

```
  lumtax(s,d)   Import duties prior to tax on Canadian exports to US
  lumflow(s,d)   Bilateral lumber trade flows for calibrating model cu m
  trans(s,d)   Bilateral shipping and handling costs per cu m

$load s data ship lumtax lumflow
$gdxin

SETS
  cs(s) Canadian supply regions /bcc, bci, alt, ont, que, mrt, pra/
  us(s) US supply regions /mw, ne, se, sw, we/

;
alias(us,ud); #US demand regions
alias(cs,cd); #Canadian demand regions
;

* -----------------------------------------------------------
* DETERMINING REGIONAL QUOTA
* -----------------------------------------------------------
PARAMETERS
  flows(s,d)   bilateral trade flows in cu m
  S0(s)   Reference converted supply quantity in region s
  D0(d)   Reference converted demand quantity in region d
  QCan   Total Canadian quota for exports to US
  Qbci   Quota for BC interior
  Qbcc   Quota for BC coast
  Qalt   Quota for Alberta
  Qpra   Quota for pra
  Qque   Quota for que
  Qont   Quota for ont
  Qmrt   Quota for mrt
  Q(s,d)   quota levels
;

trans(s,d) = ship(s,d);
flows(s,d) = lumflow(s,d)/conv;
S0(s) = data(s,'supply')/conv;
D0(d) = data(d,'demand')/conv;
Qcan = Qpercent * UScon;
Qbcc = 0.062332*QCan;
Qbci = 0.457098*Qcan;
Qalt = 0.083114*Qcan;
Qque = 0.204535*Qcan;
Qont = 0.084577*Qcan;
Qpra = 0.016753*Qcan;
Qmrt = 0.091591*Qcan;
Q(s,d) = 999999999;
```

```
Q('bci',ud) = (1/conv)*Qbci*lumflow('bci',ud)/(lumflow('bci','mw') + lumflow('bci','ne') + lumflow('bci','se') + lumflow('bci','sw')+lumflow('bci','we'));

Q('bcc',ud) =(1/conv)*Qbcc*lumflow('bcc',ud)/(lumflow('bcc','mw') + lumflow('bcc','ne') + lumflow('bcc','se') + lumflow('bcc','sw') + lumflow('bcc','we'));

Q('alt',ud)= (1/conv)*Qalt*lumflow('alt',ud) / (lumflow('alt','mw') + lumflow('alt','ne') + lumflow('alt','se') + lumflow('alt','sw') + lumflow('alt','we'));

Q('que',ud) =(1/conv)*Qque*lumflow('que',ud) / (lumflow('que','mw') + lumflow('que','ne') + lumflow('que','se') + lumflow('que','sw')+lumflow('que','we'));

Q('ont',ud) =(1/conv)*Qont*lumflow('ont',ud) / (lumflow('ont','mw') + lumflow('ont','ne') + lumflow('ont','se') + lumflow('ont','sw')+lumflow('ont','we'));

Q('pra',ud) =(1/conv)*Qpra*lumflow('pra',ud) / (lumflow('pra','mw') + lumflow('pra','ne') + lumflow('pra','se') + lumflow('pra','sw')+lumflow('pra','we'));

* ------------------------------------------------------------
* DETERMINING INTERCEPTS AND SLOPES FOR DEMAND AND SUPPLY
* ------------------------------------------------------------
PARAMETERS
  x0(s,d)      Reference bilateral trade quantity sold by region d to s
  P0(s)      Reference price in region s
  eta(s)      Price elasticity of supply
  sigma(d)      Price elasticity of demand
  dint(d)      demand intercept
  dslope(d)      demand slope
  sint(s)      supply intercept
  sslope(s)      supply slope
  prices(d)      final demand prices in each region
  tradeflow(s,d)      trade flow from calibration
  ;

  x0(s,d) = lumflow(s,d)/conv;
  P0(s) = data(s, 'price')*conv; # $ per cu m
  eta(s) = data(s,'selast');
  sigma(d) = data(d,'delast');
  dslope(d) = (-1/sigma(d))*(P0(d)/D0(d));
  dint(d) = P0(d) + dslope(d)*D0(d) ;
  sslope(s) = (1/eta(s))*(P0(s)/S0(s)) ;
  sint(s) = P0(s) - sslope(s)*S0(s) ;

* ------------------------------------------------------
* DEFINE VARIABLES CREATED AFTER MODEL SOLUTIONS
* ------------------------------------------------------
PARAMETERS
CheckDem(d)      Should equal zero
CheckSup(s)      Should equal zero
TransCost      Total transportation costs {$ millions}
UnderQuota(cs,ud) Under quota trade flows
```

```
OverQuota(cs,ud) Over quota trade flows
TotTrade(cs,ud) Total trade flows
RentQ(cs,ud)      quota rent from regions constrained by surge trigger
TarRev(cs)      SLA tariff revenue by Canadian region
GlobalTrade(s,d)      global bilateral trade after solving model;
*-------------------------------------------------------
* LUMBER QUOTA MODEL
*-------------------------------------------------------
POSITIVE VARIABLES
x(s,d)  bilateral trade flows
pd(d)      demand price
ps(s)     supply price
P(s)     regional price
QR(s,d)      quota rents
;

EQUATIONS
market(s)      Domestic market clearing constraint
demand(d)      Consumption cannot exceed domestic production plus imports
supply(s)      Production must equal or exceed domestic cons plus exports
zprofit(s,d)      Ensures markets clear to get zero profit
quota(s,d)      Trade must comply with SLA surge triggers
*Xportrow(s,d)      Calibration exports to rest of world by region
;

demand(d).. sum(s, x(s,d)) =G= (dint(d)-pd(d))/dslope(d);

supply(s).. (ps(s) - sint(s))/sslope(s) =G= sum(d,x(s,d));

quota(s,d).. Q(s,d)=G= x(s,d);

zprofit(s,d).. (1 + lumtax(s,d))*ps(s) + trans(s,d) + QR(s,d)=G= pd(d) ;

MODEL lumtrade /demand.pd, supply.ps, zprofit.x, quota.QR/;

pd.lo(d) = 0.001;
ps.lo(s) = 0.001;
pd.l(d) = P0(d);
ps.l(s) = P0(s);
x.l(s,d) = lumflow(s,d);
pd.lo("we")=170.34;

OPTION MCP = NLPEC; # Other solvers are MILES en PATH
SOLVE lumtrade using mcp;

PARAMETERS
CheckDem(d)      Should equal zero
CheckSup(s)      Should equal zero
TransCost      Total transportation costs {$ millions}
UnderQuota(cs,ud) Under quota trade flows
OverQuota(cs,ud)   Over quota trade flows
```

```
TotTrade(cs,ud)      Total trade flows
RentQ(cs,ud)      quota rent from regions constrained by surge trigger
TarRev(cs)      SLA tariff revenue by Canadian region
GlobalTrade(s,d)      Global bilateral trade after solving model
ConSur(d)      Consumer TRADE surplus (gain from trade)
ProdSur(s)      Producer TRADE surplus (gain from trade)
;

CheckDem(d) = (dint(d)-pd.l(d))/dslope(d) - sum(s, x.l(s,d));
CheckSup(s) = sum(d, x.l(s,d)) - (ps.l(s) - sint(s))/sslope(s);
TransCost = sum((s,d), trans(s,d)*x.l(s,d));
ConSur(d) = 0.5*(dint(d)-pd.l(d))* sum(d, x.l(s,d));
ProdSur(s)= 0.5* (ps.l(s) - sint(s))* sum(s, x.l(s,d));
UnderQuota(cs,ud)=x.l(cs,ud);
RentQ(cs,ud) = QR.l(cs,ud)*x.l(cs,ud);
GlobalTrade(s,d) = x.l(s,d);
prices(d) = pd.l(d);

DISPLAY GlobalTrade, RentQ, QR.l, ConSur, ProdSur;

SET prov province id /BC, Alberta, Ontario, Quebec, Prairies, Maritimes/ ;

PARAMETER provrent(prov) provincial rent ;

provrent('BC') = sum(us, RentQ('bcc', us) + RentQ('bci', us));
provrent('Alberta') = sum(us, RentQ('alt', us));
provrent('Quebec') = sum(us, RentQ('que', us));
provrent('Ontario') = sum(us, RentQ('ont', us));
provrent('Prairies') = sum(us, RentQ('pra', us));
provrent('Maritimes') = sum(us, RentQ('mrt', us));

$libinclude xldump provrent LumberHQ.xls Surge!a1
$libinclude xldump pd LumberHQ.xls Surge!a10
$libinclude xldump RentQ LumberHQ.xls Surge!a15
$libinclude xldump ConSur LumberHQ.xls Surge!a25
$libinclude xldump ProdSur LumberHQ.xls Surge!a30
$libinclude xldump GlobalTrade LumberHQ.xls Trade!a2
$libinclude xldump dint LumberHQ.xls EDES!a1
$libinclude xldump dslope LumberHQ.xls EDES!a3
$libinclude xldump sint LumberHQ.xls EDES!a6
$libinclude xldump sslope LumberHQ.xls EDES!a9
```

References

BC Statistics (2019) BC statistical series. https://www2.gov.bc.ca/gov/content/data/statistics (accessed November 7, 2019).

Comtrade (2019) UN comtrade database. Available at: https://comtrade.un.org/ (accessed November 7, 2019).

European Commission (2013) A new EU forest strategy: forests and the forest-based sector. Communication from the Commission to the European Parliament, the Council, the European Economic and Social Committee and the Committee of the Regions, 20 September. *Com(2013) 659 Final*. Brussels, Belgium.

European Commission (2014) A policy framework for climate and energy in the period from 2020 to 2030. Communication from the Commission to the European Parliament, the Council, the European Economic and Social Committee and the Committee of the Regions, 22 January. *Com(2014) 015 Final*. Brussels, Belgium.

FAO (2020) FAO forestry statistics. Food and Agricultural Organization (FAO) of the United Nations, Rome, Italy. Available at: http://www.fao.org/faostat/en/#data (accessed July 20, 2020).

Government of Canada (2019) National forestry database. Updated 16 April. Available at: http://nfdp.ccfm.org/index_e.php (accessed October 16, 2019).

Johnston, C.M.T. and Parajuli, R. (2017) What's next in the U.S.–Canada softwood lumber dispute? An economic analysis of restrictive trade policy measures. *Forest Policy and Economics* 85(1), 135–146.

Johnston, C.M.T. and van Kooten, G.C. (2014) Modeling bi-lateral forest product trade flows: experiencing vertical and horizontal chain optimization. *REPA Working Paper #2014-04*. Resource Economics and Policy Analysis Group, University of Victoria, Canada.

Johnston, C.M.T. and van Kooten, G.C. (2015) Back to the past: burning wood to save the globe. *Ecological Economics* 120, 185–193.

Johnston, C.M.T. and van Kooten, G.C. (2016) Global trade impacts of increasing Europe's bioenergy demand. *Journal of Forest Economics* 23, 27–44.

Johnston, C.M.T. and van Kooten, G.C. (2017) Impact of inefficient quota allocation under the Canada–U.S. softwood lumber dispute: a calibrated mixed complementarity approach. *Forest Policy and Economics* 74, 71–80.

Nicholson, C.F. and Bishop, P.M. (2004) U.S. dairy product trade: modeling approaches and the impact of new product formulations. *Final Report for NRI Grant #2001-35400-10249, March 2004*.

U.S. Bureau of Economic Analysis (2019) SQGDP2 GDP by state 1/ GDP by state: all industry total. Available at: https://apps.bea.gov/itable/iTable.cfm?ReqID=70&step=1#reqid=70&-step=1&isuri=1 (accessed July 20, 2020).

U.S. International Trade Commission (2017) Softwood lumber products from Canada. *Investigation Nos. 701-TA-566 and 731-TA-1342 (Preliminary), Publication 4663, January*. Available at: https://www.usitc.gov/publications/701_731/pub4663.pdf (accessed November 7, 2019).

U.S. International Trade Commission (2019) Softwood lumber from Canada. Available at: https://www.usitc.gov/investigations/701731/2017/softwood_lumber_canada/final.htm (accessed November 7, 2019).

van Kooten, G.C. (2002) Economic analysis of the Canada–United States softwood lumber dispute: playing the quota game. *Forest Science* 48(November), 712–721.

van Kooten, G.C. (2014) The benefits of impeding free trade: revisiting British Columbia's restrictions on log exports. *Journal of Forest Economics* 20(4), 333–347.

van Kooten, G.C. (2018) The challenge of mitigating climate change through forestry activities: what are the rules of the game? *Ecological Economics* 146, 35–43.

van Kooten, G.C. (2020). *Applied Economics, Trade and Agricultural Policy*. University of Toronto Press, Toronto, Canada (in press).

van Kooten, G.C. and Johnston, C.M.T. (2016) The economics of forest carbon offsets. *Annual Review of Resource Economics* 8(1), 227–246.

6 Global Forest Products Trade Model

Prakash Nepal[1], Joseph Buongiorno[2], Craig M.T. Johnston[3], Jeffrey Prestemon[4] and Jinggang Guo[5]

[1]*USDA Forest Products Laboratory, Madison, Wisconsin, USA;* [2]*University of Wisconsin-Madison, Madison, Wisconsin, USA;* [3]*Bank of Canada, Ottawa, ON Canada;* [4]*USDA Forest Service, Southern Research Station, Research Triangle Park, North Carolina, USA;* [5]*RTI International, Research Triangle Park, NC, USA*

Abstract

This chapter introduces the Global Forest Products Model (GFPM). The general model structure and the mathematical formulation of the GFPM are provided and key differences and similarities to the modeling approaches developed in the previous chapters are highlighted. The usefulness of the GFPM as a forest sector tool for policy analysis is illustrated by summarizing its applications in a wide array of past and ongoing studies. These studies are summarized under four representative groups: (i) forest sector outlook studies; (ii) studies evaluating the consequences of tariff and non-tariff barriers on the international trade of forest products; (iii) studies projecting the impacts of climate change and forest-based climate change mitigation strategies on forests and forest industries; and (iv) other studies dealing with other important questions, such as the effects of the rise in global planted forest area, illegal harvests, and invasive species. Some of the limitations of GFPM, ways to mitigate these limitations, and its overall usefulness as a forest sector policy analysis tool are also examined.

The Global Forest Products Model (GFPM) offers an alternative approach to the spatial price equilibrium (SPE) trade models described in the previous chapters and the models reviewed by Latta *et al.* (2013). The GFPM is also an SPE trade model and therefore shares fundamentally similar model constructs to the Resource Economics and Policy Analysis (REPA) model. It also utilizes demand and supply equations and data on production, consumption, prices, manufacturing costs, and transport costs to maximize the sum of the consumer and producer surpluses for all products in all regions. However, the two models differ in many respects (Table 6.1) including the following.

- GFPM is calibrated using historical goal programming as opposed to positive mathematical programming in the REPA model (see Chapter 4, section 4.2, this volume).

DOI: 10.1079/9781789248234.0006

Table 6.1. Major differences between the REPA trade model and GFPM. (Authors' own table.)

	REPA	GFPM
Mathematical optimization	QCP/MCP solved by GAMS	QCP solved by BPMPD optimizer
Time dependency	Static model	Recursive dynamic model
Geographic coverage	Global, 20 regions	Global, 180 countries
Model calibration	Positive mathematical programming	Goal programming
Sector coverage	Eight forest products	Fourteen principal categories of forest products
International trade	Bilateral trade flows at detailed product level	Trade between country and the rest of the world[a]
Forest inventory dynamics	NA	Growth of forest inventory is a function of stocking density. Forest inventory evolves over time as previous year inventory plus projected current year growth minus harvest quantity
Forest area dynamics	NA	Rate of change of forest area in each country is projected with an environmental Kuznets curve
Base year	2016	2017 with a 3-year data average (2016, 2017, 2018)

[a] Further data are needed to model the bilateral trade flow. See Buongiorno and Zhu (2018a).

- GFPM incorporates 180 countries, while the number of regions in the REPA model is flexible and has included up to 20 regions (Chapter 4). Some regions in the REPA model are at the sub-national level, and the number of regions in the model can be expanded.
- The REPA model determines bilateral trade flows between regions, whereas individual countries trade with the aggregate world market in the standard database of the GFPM. However, GFPM can also use data sets with bilateral trade flows, as described in Buongiorno and Zhu (2018a) and as applied in Turner *et al.* (2001). An R program for integrating several countries into regions and then determining bilateral trade between the regions is provided in Appendix 4.D of Chapter 4.
- There is a difference in the number of wood products that are modeled, with 14 products in GFPM versus up to eight products in the REPA model.
- The temporal dimensions differ between the models; GFPM employs dynamic recursive solutions versus static solutions in the REPA models.
- Changes in forest resources, such as forest area and forest stock, are endogenously determined in GFPM together with endogenously projected harvests; in the REPA model, a supply function for logs is specified for each region.

Finally, the GFPM was originally developed at the University of Wisconsin-Madison during the 1990s and has subsequently been maintained and enhanced.

The objective of this chapter, however, is not to compare and contrast the GFPM with REPA models, but to describe the general model structure and the mathematical formulation of the GFPM. The objective is to introduce some representative published studies illustrating its application, as a policy tool, in evaluating different questions related to forests and the forest products sector at the national, regional, and global level so that readers interested in forest products modeling can build upon the materials, descriptions, examples, and references provided here.

In this chapter, we first provide a brief history of GFPM, followed by an overview of the GFPM's general structure. This is followed by a discussion of the data and calibration, and the validation techniques used in the GFPM, along with a detailed mathematical formulation of the model. We then highlight selected past applications of GFPM in evaluating the consequences for forest resources, forest products production, consumption, trade, and prices of various scenarios of economic and demographic changes; climate change; and changes in trade and environmental policies affecting the forest sector. We conclude by noting some of the limitations of GFPM, ways to mitigate these limitations, and its usefulness as a forest sector policy analysis tool.

6.1 A Brief History of the GFPM

The first complete version of the GFPM capable of producing long-term projections of production, consumption, imports, exports, and prices by country was developed in 1996, and was applied to produce the provisional global forest products market outlook study of the Food and Agricultural Organization (FAO, 1997). Yearly updates and improvements followed, ending with the GFPM 2017 version. The GFPM 2017 software and the latest database (3-year average data for 2016, 2017, and 2018) are provided freely at http://labs.russell.wisc.edu/buongiorno/welcome/gfpm/ (accessed September 4, 2020).

The GFPM was originally built with the Price Endogenous Linear Programming System, or PELPS (Zhang *et al.*, 1993), a general software developed to model any economic sector with spatial and temporal components, coupled with the LINDO linear programming solver (Schrage, 1991). The PELPS system itself resulted from the development of a recursive-linear programming model of the paper industry (Gilless and Buongiorno, 1987; Zhang *et al.*, 1996).

Since 1996, GFPM has been continuously improved and expanded to address various national and global issues in forest economics and policy, leading to the 2017 version of the GFPM, based on QPELPS (Quadratic Price Endogenous Linear Programming System) with the interior point BPMPD program (Mészáros, 1999), making GFPM independent of commercial optimizers.

6.2 GFPM Structure

The GFPM is designed chiefly as a policy analysis tool, facilitating an understanding of how forest products production, consumption, imports, exports, prices, and welfare are likely to change under a given or a combination of scenarios of economic changes. These include changes in gross domestic products (GDP), biophysical changes (e.g. changes in forest area, growth, and inventory), changes in technology (e.g. changes in production capacity), and changes in trade and related policies (e.g. tariff and non-tariff related trade barriers) (Buongiorno *et al.*, 2003; Buongiorno, 2015). Such a capability in GFPM is enabled through the integration of the classical four major components of forest sector models (Kallio *et al.*, 1987): timber supply (production of raw materials), processing industries (manufacturing of materials into products), demand for end products, and international trade. In obtaining the market equilibrium solution in each projected year, the GFPM follows the same theory and methods as other SPE models. That is, GFPM maximizes the global 'net social payoff' – the sum of the consumer and producer surpluses for all products and countries minus transportation costs (Takayama and Judge, 1971). This is the 'static phase' of the calculations. The 'dynamic phase' refers to the yearly changes brought about by shifts in demand that change the market equilibrium conditions, for example. Thus, the model is dynamic recursive, with no inter-temporal optimization and thus free from unrealistic 'perfect foresight' assumptions. In the GFPM the future depends only on the past and on exogenous assumptions regarding macro changes, such as GDP, population or trade policies. Technically, the model is a synthesis of econometrics, mathematical programming, and system dynamic methods (Buongiorno, 1996).

The current model deals explicitly with 180 individual countries and territories (Table 6.2), including 50 countries and territories in Africa, 47 in Asia, 37 in Europe, 22 in North America, 13 in South America, and 11 in Oceania. Country-level analysis is important for at least two reasons. First, all political decisions are by individual country governments. One exception is Canada where, under the constitution, decisions related to the forest sector are made at the provincial level. Second, checking the reasonableness of model data and output is easier at the country level than at the regional level, while more people are familiar with, say, Brazil, than with 'Latin America'. The current version of the GFPM deals with 14 forest product groups (Fig. 6.1) encompassing raw materials (industrial roundwood, other industrial roundwood, fuelwood, waste paper, and other fiber pulp), intermediate products (mechanical and chemical pulp), and end products (e.g. sawnwood, plywood, particleboard, fiberboard, fuelwood, other industrial roundwood, newsprint, printing and writing paper, and other paper and paperboard). The supply of industrial roundwood, fuelwood, and other industrial roundwood is a function of its own price and forest stock, both of which are projected endogenously. The supply of waste paper (supply of recovered paper can be constrained to satisfy a specified recovery rate) and other fiber pulps (straw, bagasse, etc.) are functions of their own prices and GDP.

The demands for the nine manufactured end products modeled in the GFPM are functions of their endogenously projected own price and an exogenously projected GDP (Buongiorno, 2015). The two intermediate products, mechanical pulp and chemical pulp, are used as inputs to paper production. Demand for raw material (industrial roundwood) and intermediate products (wood pulp) are derived from the demand for end products through the input–output coefficients

Table 6.2. Countries and territories represented in the GFPM. (Authors' own table.)

Africa		Asia	
Algeria	Madagascar	Afghanistan	Malaysia
Angola	Malawi	Armenia	Mongolia
Benin	Mali	Azerbaijan	Myanmar
Botswana	Mauritania	Bahrain	Nepal
Burkina Faso	Mauritius	Bangladesh	Oman
Burundi	Morocco	Bhutan	Pakistan
Cameroon	Mozambique	Brunei Darussalam	Philippines
Cape Verde	Niger	Cambodia	Qatar
Central African Rep	Nigeria	China	Saudi Arabia
Chad	Réunion	Cyprus	Singapore
Congo, Dem Rep	Rwanda	Georgia	Sri Lanka
Congo, Rep	Sao Tome & Principe	Kazakhstan	Syria
Côte d'Ivoire	Senegal	Kyrgyzstan	Tajikistan
Djibouti	Sierra Leone	Maldives	Thailand
Egypt	Somalia	India	Timor-Leste
Equatorial Guinea	South Africa	Indonesia	Turkey
Ethiopia	Sudan	Iran	Turkmenistan
Gabon	Swaziland	Iraq	United Arab Emirates
Gambia	Tanzania	Israel	Uzbekistan
Ghana	Togo	Japan	Viet Nam
Guinea	Tunisia	Jordan	Yemen
Guinea-Bissau	Uganda	N. Korea	
Kenya	Zambia	S. Korea	
Lesotho	Zimbabwe	Kuwait	
Liberia		Laos	
Libya		Lebanon	
Europe		**North America**	
Albania	Lithuania	Bahamas	Panama
Austria	Macedonia	Barbados	St Vincent/Grenadines
Belarus	Moldova	Belize	Trinidad & Tobago
Belgium	Montenegro	Canada	USA
Bosnia–Herzegovina	Netherlands	Saint Lucia	
Bulgaria	Norway	Costa Rica	
Croatia	Poland	Cuba	
Czech Republic	Portugal	Dominica	
Denmark	Romania	Dominican Rep	
Estonia	Russian Federation	El Salvador	
Finland	Serbia	Guatemala	
France	Slovakia	Haiti	
Germany	Slovenia	Honduras	
Greece	Spain	Jamaica	
Hungary	Sweden	Martinique	
Luxembourg	Switzerland	Mexico	
Ireland	Ukraine	Netherlands Antilles	
Italy	United Kingdom	Nicaragua	
Latvia			

Table 6.2. Continued.

Oceania		South America	
Australia	Papua New Guinea	Argentina	Guyana
Cook Islands	Samoa	Bolivia	Paraguay
Fiji Islands	Solomon Islands	Brazil	Peru
French Polynesia	Tonga	Chile	Suriname
New Caledonia	Vanuatu	Colombia	Uruguay
New Zealand		Ecuador	Venezuela
		French Guiana	

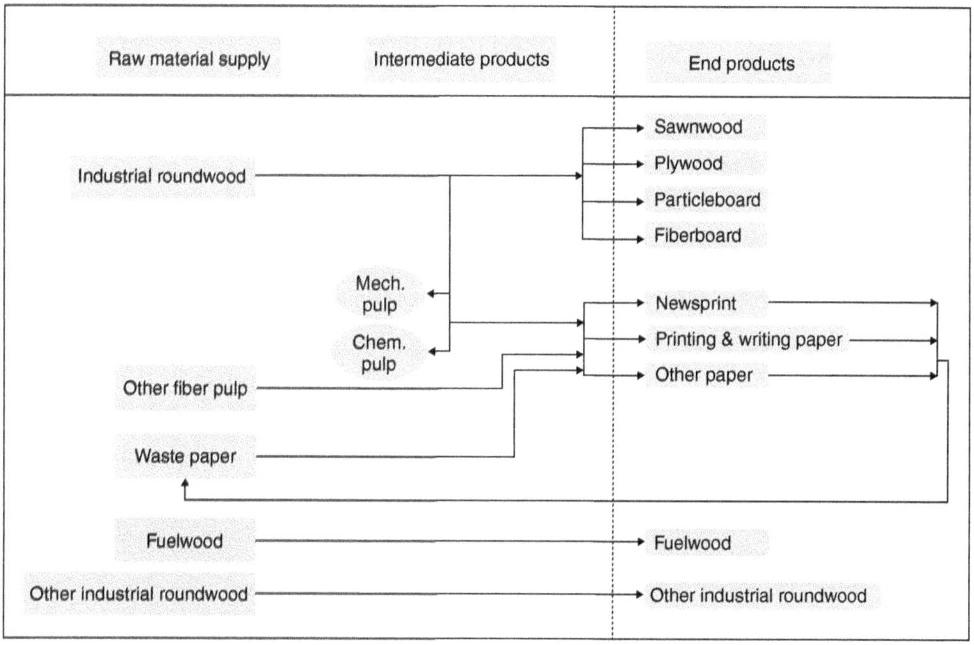

Fig. 6.1. Flow of raw materials, and intermediate and end products, as modeled in GFPM. (Authors' own figure.)

(ratio of the amount of input used in manufacturing a product to the amount of output) and manufacturing costs (labor, capital, energy). Differences in input cost, input–output efficiency, and manufacturing cost determine the comparative advantages of countries.

In the current standard data set, called the 'World file' of the GFPM, international trade (imports and exports) is between a country and the rest of the world. However, the data can be modified to represent the bilateral trade flows, as was done in Turner *et al.* (2001) and as described in Buongiorno and Zhu (2018a). Quantities of products imported or exported are driven by the competitive advantage of a country or a region in producing and shipping each product. Competitive advantage is a function of transport costs, manufacturing costs, input–output coefficients, and the endogenously determined domestic and world prices of a product. For instance, a country or a region may increase its net exports if it has a relatively more cost-effective technology (e.g. lower input requirements or lower manufacturing costs, compared with another country) in producing a particular product. In addition, changes in trade in the GFPM are limited by a priori trade inertia constraints to avoid unrealistic movements compared with historical data.

For each projected year, the model solution clears markets in all countries, obtaining prices at which demand is equal to supply for all products. The demand for end products (sawnwood, panels, pulp, and paper) in each country changes from one year to the next due to changes in GDP or up to five other exogenous demand shifters (e.g. housing construction, wood energy demand). Raw wood supply shifts according to endogenously projected changes in forest stocks or up to five other exogenously specified shifters (e.g. forest productivity due to climate change).

The forest stock shifts the supply of industrial roundwood, other industrial roundwood, and fuelwood in each country, and evolves over time as previous year stock plus projected current year growth minus harvest quantity. Forest stock growth (net of mortality) before harvest is a nonlinear function of forest stock density (forest stock per unit of forest area) based on the work by Turner *et al.* (2006). The relationship between forest growth and forest stock density implies that forest growth increases with declining stock density and decreases with increasing stock density. Changes in forest stock densities in individual countries are determined by the endogenously projected changes in forest stock and forest area. Forest area in each country is projected in the GFPM with an environmental Kuznets curve (Kuznets, 1955) originally estimated by Turner *et al.* (2006) and revised by Buongiorno (2015). The current functional form implies that forest area change is negative at low GDP per capita, becomes positive and increases at higher GDP per capita, and then decreases and approaches zero at very high GDP per capita.

Thus, the GFPM is built on the general principle that the allocation of scarce resources in the short run is optimized by global markets, while the long-run resource allocation is partly determined by the combination of market forces (e.g. projected prices) and political forces, such as changes in forest policy leading to a shift in the wood supply, changes in environmental policy affecting the wastepaper recovery, or changes in trade policies that change the cost of imports (e.g. tariff and non-tariff barriers).

6.3 Input Data, and Model Calibration and Validation

Data inputs and outputs in the GFPM are handled with Microsoft Excel spreadsheets and graphics. The GFPM needs data for the base year and the projection years. Data for the base year are mostly observed market data (e.g. quantities of production, consumption, exports and imports, and prices), biophysical data (e.g. forest stock, area), data on production technologies (e.g. input–output coefficients, manufacturing costs), and transport costs for each country, coupled with the parameters for the equations of demand, supply, manufacturing processes, and trade. Parameters of the demand and supply equations are obtained from econometric modeling (e.g. Turner *et al.*, 2006; Buongiorno, 2015). Data for the projection years are exogenously specified trajectories of demand and supply shifters or changes in other parameters.

Most of the base year data are derived from the FAOSTAT database (production, consumption, trade, and prices) and the World Bank's World Development Indicators (WDI) database (GDP and population). The data on forest stocks, forest stock growth rate, forest area, and timber growth rates are derived from the FAO's most recent Global Forest Resource Assessment. World prices are derived as the unit value of net exports (total value of net exports divided by the total quantity of net exports) as reported in the FAOSTAT database. Domestic demand and supply price of a product is equal to the world price for net exporters of the commodity, or to the world price plus freight costs and tariffs for net importers. Data on the input–output coefficients and the manufactured costs are determined simultaneously through a calibration procedure described below, which also uses the FAOSTAT data.

The purpose of the calibration procedure is to obtain a base-year model solution that matches the observed data. The calibration in the GFPM is achieved by estimating the input–output coefficients that minimize the sum of the weighted absolute deviations between estimated production and reported production, and of the sum of the weighted absolute difference between the estimated input and the input implied by prior input–output coefficients as suggested by technical knowledge (Buongiorno, 2015). The resulting input–output coefficients are used to further estimate manufacturing costs, which are equal to the unit value of the output minus the cost of all inputs evaluated at local prices. In calculating the manufacturing costs, world prices are used as the export prices facing the net exporter, while import prices are calculated as world prices plus transport costs and tariffs and used for the net importer countries. Thus, manufacturing costs plus other input costs (wood and other fibers) exactly offset output revenues, resulting in zero profits, as they should be in a competitive equilibrium.

The World data set available with the latest version of the GFPM (GFPM 2017) was calibrated with a 3-year data average (2016, 2017, 2018), for a base year of 2017 (Buongiorno and Zhu, 2018a). After calibration, the GFPM software checks for data consistency, ensuring that the following conditions hold for the base-year:

- Apparent consumption (production plus import minus export) equals final demand, or intermediate demand for input used by other products.
- Local price equals the world price plus the transport cost for net importers, or the world price for net exporters.
- Manufacturing cost equals the price of the output minus the cost of all inputs, given the price of inputs and the input–output coefficients.
- The waste paper used in paper manufacturing does not exceed the recovered waste paper, given the paper consumption and maximum recovery rate.

- This calibrated and consistent model is then tested with the GFPM validation procedure, which verifies that the model solution is close to the actually observed data in the base year.

More details on calibration and validation techniques are provided in the GFPM user manuals (Buongiorno and Zhu, 2018a,b).

6.4 Mathematical Formulation

In the following description, the static phase refers to the market optimization in any given year of the projection. The dynamic phase refers to the calculations that simulate periodic changes in market conditions, such as those due to economic or demographic growth.

6.4.1 Spatial global equilibrium (static phase)

The spatial competitive global market equilibrium of the forest products sector in a given year is obtained in the GFPM by maximizing the following quadratic objective function (Buongiorno, 2015):

$$\begin{aligned} Maximize\ Z = & \sum_i \sum_k \int_0^{D_{ik}} P_{ik}(x)dx - \sum_i \sum_k \int_0^{S_{ik}} P_{ik}(z)dz - \sum_i \sum_k \int_0^{Y_{ik}} m_{ik}(y)dy \\ & - \sum_i \sum_j \sum_k c_{ijk} T_{ijk}, \end{aligned} \tag{6.1}$$

where x, y and z are integration factors; i and j refer to countries and k refers to a product; P is price in U.S. dollars of constant value; D is final product demand; S is raw material supply; Y is quantity manufactured; m is cost of manufacture (labor, capital, and materials excluding wood and fiber); T is quantity transported; and c is freight cost (cost of transport plus tariff).

The objective function in (6.1) maximizes the total welfare in the global forest sector in a given year. It equals the surplus value of the products to consumers (consumer surplus as the area under all the demand curves above price lines) minus the cost of supplying the raw materials (quasi-rent measured as the area above the country supply curves below price).

The objective function in equation (6.1) is maximized subject to the specified constraints related to end product demand (equation (6.2)), raw material supply (equation (6.3)), wood drain (equation (6.4)), material balance (equation (6.5)), trade inertia (equation (6.6)), manufacturing costs (equation (6.7)), and transport costs (equations (6.8) and (6.9)).

Equation (6.2) specifies that the demand for each end product in each country in a given year has a constant elasticity with respect to the price, where D^* is current consumption at last period's price, P_{-1}, and δ is the price elasticity of demand. As shown in the section on the dynamic phase below, D^* depends on last period's demand, the growth of a country's GDP, and other exogenous or endogenous demand shifters.

$$D_{ik} = D^*_{ik}\left(\frac{P_{ik}}{P_{ik,-1}}\right)^{\delta_{ik}}. \tag{6.2}$$

According to equation (6.3), the supply of raw materials in each country in a given year has a constant elasticity with respect to the price, where S^* is current supply at last period's price and λ is the price elasticity of supply. As shown in the dynamic phase below, S^* depend on last period's supply and on exogenous or endogenous supply shifters, especially the forest stock.

$$S_{ik} = S^*_{ik}\left(\frac{P_{ik}}{P_{ik,-1}}\right)^{\lambda_{ik}}. \tag{6.3}$$

Equation (6.4) defines the total wood drain from the forest, where r refers to industrial roundwood, n to other industrial roundwood, f refers to fuelwood, $0 \leq \theta_i \leq 1$ is the fraction of fuelwood that comes from the forest, and $\mu \geq 1$ is the ratio of drain from forest stock to harvest:

$$S_i = (S_{ir} + S_{in} + \theta_i S_{if})\mu_i. \tag{6.4}$$

Since the wood drain cannot exceed the available stock, the following constraint applies:

$$S_i \leq I_i, \tag{6.4'}$$

where I_i is the current forest stock.

Optional constraints may limit the harvest to a fraction of the growth of forest stock, such as the 'allowable cut constraints' described in equation (6.27) below.

The next constraint defines the material balance for each country:

$$\sum_j T_{jik} + S_{ik} + Y_{ik} - D_{ik} - \sum_n a_{ikn} Y_{in} - \sum_j T_{ijk} = 0, \forall_{i,k}, \tag{6.5}$$

where a_{ikn} is the input of product k per unit of product n. According to equation (6.5), in each country and for each product, the quantity imported, the domestic supply, and the manufactured quantity must equal the domestic demand plus the quantity used in manufacturing other products, plus exports.

In addition, the production of by-products, which results from the manufacture of a commodity (e.g. sawmill residues), can be optionally specified with the following constraints:

$$Y_{il} - b_{ikl} Y_{ik} = 0, \forall_{i,k,l}, \tag{6.5'}$$

where b_{ikl} is the by-product l that can be recovered per unit of manufactured commodity k.

Equation (6.6) defines the trade inertia constraints, introduced to quantify a dynamic adjustment process in trade in response to price and income changes.

These constraints keep the current trade quantity within a lower bound (L) and an upper bound (U), relative to the previous period:

$$T_{ijk}^{L} \leq T_{ijk} \leq T_{ijk}^{U}. \tag{6.6}$$

According to equation (6.7), the manufacturing cost is a function of manufacturing quantity. Each manufacturing process is represented by activity analysis, with input–output coefficients and a manufacturing cost. As discussed in Chapter 3, the manufacturing cost is the marginal cost of the inputs not recognized explicitly by the model (labor, energy, capital, etc.):

$$m = m_{ik}^{*}\left(\frac{Y_{ik}}{Y_{ik,-1}}\right)^{s_{ik}}, \tag{6.7}$$

where m^* is the current manufacturing cost at last period's output and s is the elasticity of manufacturing cost with respect to output. As shown in the dynamic phase section below, m^* depends on last period's manufacturing cost and on the exogenous rate of change of manufacturing cost.

Equation (6.8) defines the transport cost per unit of volume for commodity k from country i to country j in any given year:

$$c_{ijk} = c_{ijk}^{*}\left(\frac{T_{ijk}}{T_{ijk,-1}}\right)^{\tau_{ijk}}, \tag{6.8}$$

where c^* is the current transport cost at last period's trade quantity and τ is the elasticity of transport cost with respect to trade quantity. As shown in the next section, c^* depends on last period's transport cost and on exogenous changes of freight rates and taxes. In the base year, c_{ijk} is computed as:

$$c_{ijk} = f_{ijk} + t_{jk}^{X}\left(P_{ik,-1}\right) + t_{jk}^{I}\left(f_{ijk} + P_{ik,-1}\right), \tag{6.9}$$

where c is the transport cost per unit of volume, f is the freight cost per unit of volume, t^X is the export tax, t^I is the import ad valorem tariff, and P_{-1} is the observed base-year world export price.

Upon satisfying all the constraints defined by equations (6.1) through (6.9), the shadow prices of the material balance constraints (equation (6.5)) give the market-clearing prices for each commodity and country.

6.4.2 Dynamic phase

Projections from the base year to future years are enabled in GFPM through calculation of the changes in the condition of the global equilibrium from one period to the next (referred to as the dynamic phase), as summarized in equations (6.10) to (6.24) (Buongiorno, 2015). Variables in these equations refer to one country, one commodity, and one year; the rate of change refers to a multi-year period unless otherwise indicated.

Periodic changes

Periodic exponential rates of change, r_p, are defined with annual exponential rates of change, r_a, and the length of a period in years, p:

$$r_p = (1+r_a)^p - 1 \tag{6.10}$$

Periodic linear changes, Δv_p, are defined by the annual linear change, Δv_a :

$$\Delta v_p = p\Delta v_a. \tag{6.11}$$

Demand shifts

From one period to the next, demand shifts with the exogenously specified rates of growth of the demand shifters:

$$D^* = D_{-1}\left(1+\alpha_y g_y + \alpha_0\right), \tag{6.12}$$

where g_y is the GDP periodic growth rate, α_y is the elasticity with respect to GDP, and α_0 is a periodic trend.

Supply shifts

The supplies of industrial roundwood, fuelwood, and other industrial roundwood shift periodically, according to the endogenous periodic rate of change of forest stock (g_I) and forest area (g_a), and with elasticities β_I and β_a:

$$S^* = S_{-1}(1+\beta_I g_I + \beta_a g_a). \tag{6.13}$$

Similarly, the supply of recovered paper and other fiber pulp shift periodically, according to the exogenously specified periodic GDP growth rate (g_y) and the corresponding elasticity (β_{Iy}):

$$S^* = S_{-1}(1+\beta_{Iy} g_y). \tag{6.14}$$

The periodic changes in forest area in an individual country are defined by:

$$A = (1+g_a)A_{-1}, \tag{6.15}$$

where A refers to the forest area and g_a to the periodic rate of change in forest area, which is based on the period length p (equation (6.10)), and the annual rate of change in forest area g_{aa}:

$$g_{aa} = (\alpha_0 + \alpha_1 y')e^{\alpha_2 y'}, \tag{6.16}$$

where y' refers to income per capita, which is predicted as a linear function of previous-period per capita income:

$$y' = (1 + g_{y'})y'_{-1}. \tag{6.17}$$

For each country, α_0 is calibrated so that in the base year the observed g_{aa} is equal to the predicted g_{aa} from equation (6.16), given the income per capita y'.

Forest stocks in individual country are projected to evolve over time according to the following growth-drain relation:

$$I = (I_{-1} + G_{-1} - pS_{-1}), \tag{6.18}$$

where I is the forest stock at the beginning of the current period and G_{-1} is the change of forest stock without harvest during the previous period.

At any point in time, the forest stock I equals $U \times A$, where A is the area and U is the stock density (stock per unit area). Without harvests, the stock growth rate is $dI/I = dU/U + dA/A$, or $g_I = g_u + g_a$. Therefore, stock growth without harvests is simply $G_{-1} = I_{-1}(g_u + g_a)$.

An optional exogenous adjustment, g_u^* may be added to represent, for example, the effects of invasive species or of climate change on the rate of stock growth, so that the final expression for stock growth is:

$$G_{-1} = I_{-1}(g_u + g_a + g_u^*). \tag{6.18'}$$

The periodic rate of forest growth without harvest, g_u, is based on the annual rate of forest growth without harvest, g_{ua}, and is defined by:

$$g_{ua} = \gamma_0 \left(\frac{I_{-1}}{A_{-1}} \right)^{\sigma}, \tag{6.19}$$

where σ is negative so that g_{ua} decreases with stock per unit area. For each country, the GFPM calibrates γ_0 automatically, so that in the base year the observed g_{ua} is equal to the g_{ua} predicted by equation (6.19) given the stock per unit area, I/A.

The periodic rate of change of forest stock net of harvest, used in equation (6.13), is then given by:

$$g_I = \frac{I - I_{-1}}{I_{-1}}. \tag{6.20}$$

Changes in input–output coefficients and manufacturing costs

The input–output coefficients, a, in the material balance constraint (equation (6.5)) can change exogenously over time (e.g. to reflect increasing use of recycled paper in paper manufacturing) as follows:

$$a = a_{-1} + \Delta a, \tag{6.21}$$

where Δa is the periodic change in the input–output coefficient.

Similarly, the manufacturing cost function (equation (6.7)) can shift exogenously over time at the annual periodic exponential rate, g_m:

$$m^* = m_{-1}(1+g_m). \tag{6.22}$$

Changes in transport cost and trade inertia bounds

The transport cost function, equation (6.8), shifts exogenously over time according to equation (6.23), by recursion of equation (6.9):

$$c^* = c_{-1} + \Delta f + t^x p_{-1} - t^x p_{-2} + t^I(f + p_{-1}) - t^I_{-1}(f_{-1} + p_{-2}), \tag{6.23}$$

where $f = f_{-1} + \Delta f$, $t = t_{-1} + \Delta t$, and Δf and Δt are periodic changes in freight costs and taxes, respectively.

Finally, the lower and upper trade inertia bounds change exogenously according to:

$$T^L = T_{-1}(1-\varepsilon)^P,\ T^U = T_{-1}(1+\varepsilon)^P, \tag{6.24}$$

where ε is the exogenously-specified absolute value of the maximum annual rate of change in trade flow.

6.4.3 Modeling timber supply with carbon markets

The GFPM allows for the modeling of timber supply when wood producers are paid for leaving trees standing to sequester carbon. This is achieved with equation (6.26), which states that the marginal cost of wood in the presence of a carbon offset payment is equal to the marginal cost of harvesting and local delivery, represented in equation (6.3), plus the opportunity cost of losing the carbon payment by not leaving the trees standing. This opportunity cost is equivalent to a shift of the supply curve commensurate with the magnitude of the carbon offset payment that is introduced (Buongiorno and Zhu, 2013).

Wood supply in the absence of a carbon market is approximated as a linear function by the tangent at the current equilibrium point (P_0, S_0):

$$P = a + bS, \tag{6.25}$$

where $b = P_0/\sigma S_0$ and $a = P_0 - bS_0$. Then, wood supply in the presence of a carbon offset payment is given by:

$$P = a + bS + \varpi(P^C - P^C_{-1}), \tag{6.26}$$

where ϖ is the CO_2 content of the forest stock (tonnes/m^3), P^C is the current price of CO_2 (\$/tCO_2), and P^C_{-1} is the price of CO_2 in the previous period.

6.4.4 Allowable cut constraint

Equation (6.27) can be optionally specified in GFPM to simulate allowable cut constraints, which are in addition to the permanent constraint that limits the total wood drain from the forest to the available stock (equation (6.4′)). The allowable cut constraint in equation (6.27) states that the drain must be less than a user-specified fraction of the current annual growth of the forest stock:

$$S \leq max\ (aG/p,0), \tag{6.27}$$

where S refers to the total wood drain from the forest defined by equation (6.4), G is the periodic change in growing stock in the absence of harvest (equation (6.19)), and a is the user-defined maximum ratio of inventory drain to the growth of growing stock.

6.5 Model Calibration and Validation

The calibration procedure of the GFPM estimates input–output coefficients and manufacturing costs based on FAO data. The procedure keeps the FAO data on exports and imports unadjusted, because traded commodities typically go through custom controls at borders for compliance with laws and regulations, and are usually more reliable than production data. However, the GFPM procedure allows for adjustments of production data for any arbitrary country in any arbitrary year as reported by FAO if such data are inconsistent with prior knowledge on the possible range of the input–output coefficients (Buongiorno, 2015).

The estimated production and input–output quantities for an individual country and year are obtained with goal programming. The objective function minimizes the sum of the weighted absolute deviations between estimated production and reported production, and of the sum of the weighted absolute differences between the estimated inputs and the inputs implied by prior input–output coefficients suggested by technical knowledge:

$$minimize\ Z = \beta \sum_{k \in A} w_k (Y_k^+ + Y_k^-) + (1-\beta) \sum_{k \in I} \sum_{n \in O} (w_k w_n)^{1/2} (Y_{kn}^+ + Y_{kn}^-). \tag{6.28}$$

In equation (6.28) and the following equations, all variables are in capital letters and the data are in lower cases. All the variables are non-negative. The data and variables refer to a specific country and year, and the subscripts k and n refer to products. In the objective function in equation (6.28), the variables Y_k^+ and Y_k^- are the deviations of estimated input k in the production of output n above or below the inputs implied by prior input–output coefficients. The weights w_k and w_n are proportional to the product prices to allow more deviation between observed and estimated products for cheap products, and $\beta = 0.90$ to give more weight to the deviations between observed and actual production rather than to deviations between estimated and expected outputs, since FAO data are available for production but no direct data are available for the inputs in a particular output.

The objective function in equation (6.28) is minimized subject to the following constraints:

$$Y_k + Y_k^- - Y_k^+ = q_k \quad \forall k \in A \tag{6.29}$$

$$Y_{kn} - \bar{a}_{kn} Y_n + Y_{kn}^- + Y_{kn}^+ = 0 \quad \forall k \in I, n \in O \tag{6.30}$$

$$Y_k \geq X_k - Z_k \quad \forall k \in F \tag{6.31}$$

$$Y_k - \sum_{n \in O} Y_{kn} = X_k - Z_k \quad \forall k \in R \tag{6.31'}$$

$$Y_{kn} - a_{kn}^U Y_n \leq 0 \quad \forall k \in I, n \in O \tag{6.32}$$

$$Y_{kn} - a_{kn}^U Y_n \leq 0 \quad \forall k \in I, n \in O \tag{6.32'}$$

$$\sum_{k \in I} Y_{kn} - a_n^U Y_n \leq 0 \quad \forall n \in O \tag{6.33}$$

$$\sum_{k \in I} Y_{kn} - a_n^L Y_n \geq 0 \quad \forall n \in O \tag{6.33'}$$

$$Y_k - \sum_{n \in F} r_{kn}^U Y_n \leq \sum_{n \in F} (Z_n - X_n) r_{kn}^U Y_n \quad \forall k \in E \tag{6.34}$$

$$Y_k - \sum_{n \in F} r_{kn}^L Y_n \geq \sum_{n \in F} (Z_n - X_n) r_{kn}^L Y_n \quad \forall k \in E \tag{6.34'}$$

$$p_k Y_k - \sum_{n \in I} Y_{nk} p_n \leq m_k^U \quad \forall k \in O \tag{6.35}$$

$$p_k Y_k - \sum_{n \in I} Y_{nk} p_n \geq m_k^L \quad \forall k \in O \tag{6.35'}$$

Equation (6.29) defines the deviation of estimated production (Y_k) from the reported production in FAOSTAT (q_k), and A is the set of products.

Equation (6.30) defines the deviations (Y_{kn}^-, Y_{kn}^+) of the estimated input of k in product n (Y_{kn}), above or below the input expected from the prior input–output coefficients ($\bar{a}_{kn} = \frac{1}{2}\left(a_{kn}^L + a_{kn}^U\right); a_{kn}^L$ and a_{kn}^U are the lower and the upper bounds on input k per unit of output n; and I and O are the respective sets of inputs and outputs.

Constraints (6.31) and (6.31′) specify that the apparent consumption of the end products must be non-negative and that an exact equality must hold for raw materials or intermediate products used in making other products, as in constraint (6.31′). X_k and Z_k are reported imports and exported quantity, respectively, which are assumed to be error-free. F is the set of end products and R is the set of raw materials or intermediate products.

Constraints (6.32) and (6.32′) limit the estimated input–output coefficients for solidwood products (e.g. quantity of industrial roundwood per unit of sawnwood) between the prior lower and upper bounds (a_{kn}^L, a_{kn}^U), as suggested by engineering knowledge. Similarly, constraints (6.33) and (6.33′) force the estimated input–output coefficients for multiple inputs (e.g. tons of mechanical pulp, chemical pulp, other fiber pulp, and waste paper per ton of newsprint) to lie between the prior technical lower and upper limits (a_n^L, a_n^U).

Constraints (6.34) and (6.34′) specify the respective upper and the lower bounds on the recovery rates of recycled product k (e.g. waste paper) from product n (e.g. newsprint). E is the set of recycled products. Lastly, constraints (6.35) and (6.35′) specify the upper and lower bounds on the unit manufacturing costs. Variables p_k and p_n are the world prices (unit values of world exports) for net exporters and the world prices plus the transport costs and tariffs for net importers, respectively.

After solving the problem specified by equations (6.28) through (6.35), the estimated input–output coefficients (the amount of product k used in making product n) are given by:

$$\hat{a}_{kn} = \frac{Y_{kn}}{Y_n} \quad \forall k \in I, n \in O. \tag{6.36}$$

The estimated input–output coefficients are used further to estimate the manufacturing costs:

$$\hat{m}_k = P_k - \sum_{n \in I} \hat{a}_{kn} P_n \quad \forall k \in O. \tag{6.37}$$

Equation (6.37) assumes a competitive market equilibrium with zero net profit so that the manufacturing cost (cost of labor and materials excluding wood and fiber and a normal return to capital) is equal to the price of the output minus the cost of wood and fiber input. Such an assumption is necessary, due to a lack of manufacturing cost data in the forest industries of most countries.

This calibration procedure is usually performed for the base year only. However, it can be replicated for earlier years to detect trends in input–output coefficients (technical change) and in manufacturing costs (Buongiorno and Zhu, 2015a). These trends are included in the GFPM 2017.

6.6 Computer Software

The GFPM integrates three different kinds of software: the Microsoft Excel spreadsheet and graphics to handle data inputs and outputs, QPELPS to set up the static and dynamic phases of the GFPM, and BPMPD (Mészáros, 1999) to calculate equilibrium in each year. The QPELPS, a general economic modeling system, takes data on the current sector state and predicted exogenous changes from a spreadsheet, writes a quadratic programming matrix expressing the GFPM static phase in MPS format, invokes the BPMPD interior point optimizer (Mészáros, 1999) to find the current year global equilibrium, updates the data to reflect this last solution and exogenous changes (dynamic phase), and starts the process again for the next period, until the time horizon is reached by successive iterations of data updating and optimization.

In the static phase, the demand, supply, and cost functions (equations (6.2), (6.3), (6.7), and (6.8)) are linearly approximated by their tangents at the current equilibrium point. Therefore, the equilibrium problem in any given year is quadratic in the objective function and linear in the constraints, and solved with BPMPD. The more time-consuming part of each periodic iteration is the data reading and updating of the dynamic phase. The most recent version of the software, the data, and the user and calibration guides are described in Buongiorno and Zhu (2018a,b) and are freely available for academic research at http://labs.russell.wisc.edu/buongiorno/welcome/gfpm/ (accessed September 4, 2020).

6.7 GFPM Application: Some Examples

With a history of more than two decades of continuous evolvement, the GFPM has found applications in numerous studies related to the national, regional, and global forest sector. These studies, which utilize the original GFPM (Buongiorno *et al.*, 2003) or its later extended versions, include several forest sector outlook reports that paint a broad picture of how the forest sector would look in the future. These projections reflect different visions of economic and demographic changes, and they typically involve scenarios of structural changes in demand or supply factors. Demand changes might include simulated shifts in the demand for particular categories of manufactured forest products (e.g. simulating the effects of more use of wood in new high-rise building construction, a decline in the demand for newsprint and printing and writing paper due to increasing internet use, etc.). Other studies deal with shocks in the supply of raw materials (e.g. due to natural catastrophes such as insect and disease outbreaks, changes in forest productivity due to climate change and CO_2 fertilization). More applications evaluate the impacts of policy-induced changes in demand and supply and forest products trade; applications include, for example, expanded use of wood for energy, increases in afforestation or reforestation to mitigate climate change, implementation of tariff or non-tariff trade barriers, and the introduction of carbon offset payment programs to encourage delayed or reduced timber harvests.

Describing all those studies is beyond the scope of this chapter. However, we present some representative examples, illustrating how the GFPM has been or is being successfully utilized to evaluate the impacts of important policy questions affecting the national, regional, and global forests and the industrial sectors that depend on them. We group these examples into four categories illustrating the wide range of applications of the GFPM: (1) long-range forest sector outlook studies; (2) assessments of the impacts of trade policies; (3) investigation of the consequences of policies aimed at climate change mitigation, including impacts of climate change on forest productivity and their effect on forest industries; and (4) miscellaneous studies.

6.7.1 Applications in outlook studies

The first application of GFPM took place in 1996 at the University of Wisconsin-Madison, on commission from the FAO. The work was done with a modified version of the PELPS system (Zhang *et al.*, 1993) and resulted in FAO's provisional outlook study (FAO, 1997), later updated in the FAO's 1999 Global Forest Products Outlook (Zhu *et al.*, 1998). These studies provided theoretically consistent projections for 180 countries in the FAOSTAT database of forest products consumption, production, trade, and prices from 1995 to 2010. Projections were based on country-specific exogenous GDP growth rates. Wood supply was price dependent, but the shifts of the wood supply functions over time were still exogenous at the time. Distinct elasticities of demand with respect to GDP and prices were estimated econometrically for high- and low-income country groups. No systematic calibration procedure had yet been found, and the input–output coefficients and manufacturing costs were estimated by Antti Rytkonen.

The GFPM was extensively used later as part of the United States Department of Agriculture (USDA) Forest Service's 2010 RPA Assessment studies, mandated by the Forest and Rangeland Renewable Resources Planning Act (RPA) of 1974. The Act requires that the U.S. Forest Service develop nationwide assessments of forest resource demand, supply and forest resource conditions every 10 years, updated every 5 years. The 2010 RPA Assessment integrated several models to project the future of forests and their industries. Two of these models were a stand-alone GFPM (Buongiorno, 2015) and its modified version, the U.S. Forest Products Model (USFPM)/GFPM (Ince *et al.*, 2011), which provides more detailed representation of wood product markets within the United States (Fig. 6.2). The USFPM/GFPM uses additional U.S. forest product demands, such as softwood lumber and hardwood lumber in contrast to a single sawnwood category in the GFPM. It also introduces more detail in U.S. forest product production, timber harvests, and timber stumpage markets (e.g. sawtimber and non-sawtimber

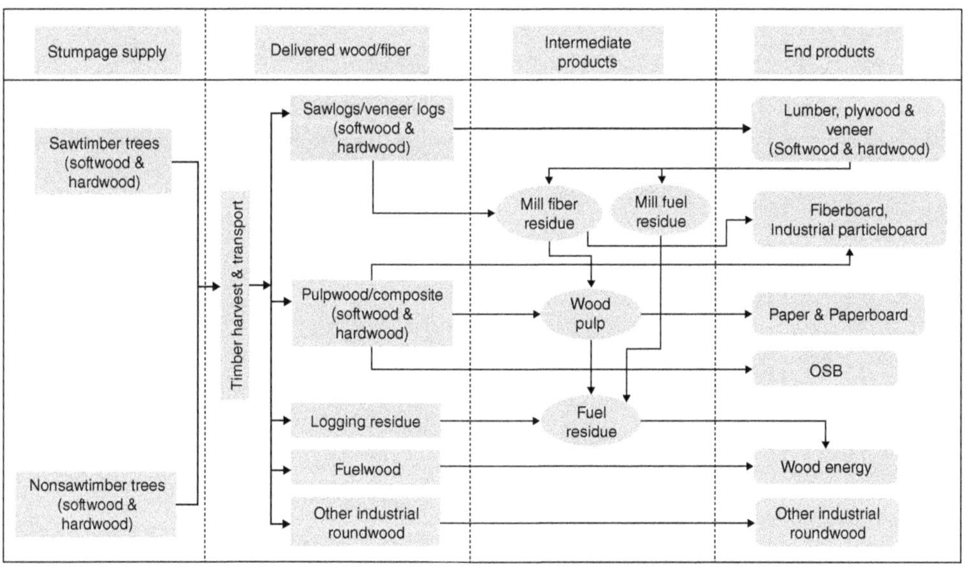

Fig. 6.2. Processes and connections among timber, delivered wood and fiber, intermediate products, and end products representing U.S. markets modeled in the modified GFPM. This modified version of GFPM is referred to as USFPM/GFPM.

harvests by hardwoods and softwood species group) for three U.S. subregions: North, South, and West. The USFPM/GFPM also simulates production of different forest-based wood biomass feedstock to meet the future increases in U.S. demand for wood energy, including roundwood fuelwood and logging and mill residues. Also important is the ability to more directly model the effects of U.S. housing construction on U.S. softwood lumber and structural and non-structural panels demand, and the effects of the displacement of graphics paper by electronic media.

One of the key policy questions in the 2010 RPA studies was how future expansion in wood energy consumption and changes in income and population would affect the forest sector in the United States and in the rest of the world. Both the standalone GFPM (Raunikar *et al.*, 2010; Buongiorno *et al.*, 2011) and the USFPM/GFPM (Ince *et al.*, 2011) were used to that end. The underlying scenarios – in particular, future changes in GDP and population – were developed in conjunction with the 4th assessment studies of the Intergovernmental Panel on Climate Change (IPCC) (Nakicenovic *et al.*, 2000) and the 2010 RPA studies (USDA Forest Service, 2012).

The approach consisted of comparing simulated projections for a reference and an alternative scenario, both of which shared exactly the same assumptions about economic drivers but differed in the projected levels of wood energy consumption. Thus, the differences in the projected outcomes between the two scenarios showed the effects on traditional forest products markets of the future changes in quantities of wood energy produced and consumed. While both the stand-alone GFPM and the USFPM/GFPM provided similar general conclusions regarding the global effects of increased wood energy (i.e. increased roundwood prices, increased lumber production, and reduced timber inventory), the results differed markedly for the U.S., mainly because of the addition of logging and mill residues in USFPM/GFPM, which satisfied the target demand for wood for energy, in contrast to the GFPM projections, which used harvested roundwood only as the wood energy source.

Another study by Ince and Nepal (2012), carried out for the 2015 Update of the 2010 RPA, employed the USFPM/GFPM to evaluate the long-term effects (to 2060) of recent trends and structural changes in U.S. forest product markets. The study included the 2007–2009 recession and recovery, a projected weakening of the U.S. dollar, and a modest increase in wood energy demand in line with historical growth rates.

In one of its most recent applications, the standalone GFPM is the main modeling tool for the 2020 UNECE Forest Sector Outlook Study (FSOS III) (Nepal and Prestemon, 2019). (The final report is expected in early 2021.) The scenarios being evaluated include five different shared socioeconomic pathways (SSPs) developed in conjunction with the fifth climate change assessment studies led by the IPCC (Riahi *et al.*, 2017). These SSPs represent alternative world visions that narratively and quantitatively describe future global changes in income, population, technology, energy use, and land-use with a varying degree of challenges for climate change mitigation and adaptation (Riahi *et al.*, 2017), consistent with global greenhouse gas and aerosol concentration trends referred to as the representative concentration pathways, or RCPs (IPCC, 2018).

The economic and demographic drivers shaping the reference scenarios adopted for the FSOS III include GDP and population (IIASA, 2018), and forest area (total forest and planted forest) projected under each SSP, by country. The projections of total forests area (Nepal *et al.*, 2019a) and planted forest area (Korhonen *et al.*, 2020) in each country were driven by GDP per capita and by projected rural population density and labor growth under each SSP.

Paired with the reference SSPs, different sets of alternative or policy scenarios are being evaluated to address three broad policy issues: (i) the sectoral effects of forest productivity change due to climate change; (ii) effects of climate change mitigation efforts; and (iii) the consequences of assumed structural changes in forest products markets. To evaluate the impact of climate change, exogenously projected changes in net primary productivity (NPP) are inserted in the GFPM, with parameter g_u^* in equation (6.18′), whereby the endogenous growth in forest stock in each of the 180 countries is adjusted by the projected change in NPP provided by the dynamic global vegetation model MC2 (Kim *et al.*, 2017). The projected forest and market outcomes from this run are then compared with a reference run, where no such adjustment was made to the endogenous forest growth stock. Because all other economic drivers are the same in the two runs, the differences in forest sector projections in these two runs are attributed to the changes in forest productivity induced by climate change. Similarly, the effects of climate change mitigation policies are projected by exogenously adjusting the growth rate of forest area (parameter g_{aa} in equation (6.16)) above the levels projected in the reference scenario. Lastly, the impacts of assumed future structural changes in wood products demand are simulated by exogenously shifting the demand for sawnwood, panels and paper products in the GFPM until the assumed target increases in demand are achieved in all countries.

6.7.2 Application in studies evaluating impacts of trade policies

The GFPM has also been applied in several studies of the effects of trade policies, such as multilateral trade agreements or country-specific tariffs, on the global trade of forest products, their production, their demand, and their prices. While early studies dealt with trade liberalization (Zhu *et al.*, 2001), recent ones are more about trade restrictions. For example, Zhu *et al.* (2008) deals with the impact of countervailing duties on coated-free sheet paper imported by the U.S. from China, the Republic of Korea, and Indonesia. The U.S. sees some increases in producer revenues, consumer expenditures, and value added, but they are small compared with the gains in Canada.

Buongiorno and Johnston (2018a) used the GFPM to estimate the potential impact of prohibitive import barriers in the United States and of countervailing measures abroad on the economic welfare of consumers and producers in the forest sectors of the United States and of other countries. The approach consists in comparing the current state of the world with two GFPM-generated scenarios, one without U.S. imports of forest products (to represent the case of prohibitive import tariffs in the U.S.), and another without U.S. imports and also without U.S. exports (to represent the case of countervailing measures by other countries). The results suggest that implementation of prohibitive import tariffs without countervailing measures by foreign countries would result in increased U.S. producer surplus, but that such gains would not be enough to offset the loss in U.S. consumer surplus. In contrast, such measures would result in increased consumer surplus in the rest of the world, but would not be enough to offset the estimated loss in producer surplus in the rest of the world. The introduction by foreign countries of countervailing measures against U.S. exports results in increased U.S. consumer surplus at the expense of greater loss in U.S. producer surplus. For the rest of the world, such a measure increases the foreign producers' surplus, but by less than it lowers their consumers' surplus. Even without foreign retaliation, imposition of prohibitive trade barriers on U.S. forest product imports hurts U.S. consumers of forest products more than it benefits its producers, leading to a net welfare loss. A trade war, with countervailing measures against U.S. exports, further decreases welfare in the United States and abroad. Similar results were also noted in Chapter 5 using the REPA trade model.

Buongiorno *et al.* (2017) is another example of GFPM application to assess the gains and losses in the forest sector resulting from international trade. In this case, the GFPM is run in comparative-statics mode. One run of the GFPM replicates a business-as-usual situation with current trade restrictions. Another run simulates pure autarky in all countries by constraining exports and imports to zero. The results suggest that unrestricted free trade among groups of countries leads to increased global economic welfare of the forest products sector, but with unequal distribution of gains and losses in welfare, especially between high-income and low-income countries. While wood producers in developed countries increase their profits with trade, those in developing countries incur heavy losses that negate any incentive to invest in forest conservation, management and new plantations.

The impact of international trade agreements has also been the subject of GFPM applications. For example, Buongiorno *et al.* (2014) investigated the effects of the transatlantic trade and investment partnership (TTIP) on the global forest sector, based on the macroeconomic impacts estimated by other authors. Comprehensive tariff elimination per se has little effect on the forest sector. But with deeper reforms and integration, consumption increases twice as much in percentage terms in the U.S. as in the EU. Net trade decreases in the U.S. more than in the EU, while it increases in Asia. The welfare of consumers and producers increases by $7 billion in the EU and $14 billion in the U.S., but decreases in some other countries, especially in Asia.

A similar GFPM approach was used in Buongiorno and Zhu (2017) to assess the potential effects of the Trans-Pacific Partnership (TPP) potential agreement between Australia, Canada, Chile, Brunei Darussalam, Japan, Malaysia, Mexico, New Zealand, Peru, Singapore, the United States, and Vietnam (TPP12), and of an extension to include India, China, and the Republic of Korea (TPP15). The comparison with and without TTP was based on changes in GDP growth projected by macroeconomic studies. The results show that with TPP12, the main total welfare gains are in the United States and Vietnam, and the greatest losses in China and Korea. Global welfare gains are larger under TPP15 to the advantage of China, Japan, and Korea, but to the detriment of the United States, the European Union and the rest of the world.

The GFPM was also applied to forecast the effects of Brexit on the global forest products industry (Johnston and Buongiorno, 2017). One optimistic and one pessimistic scenario were used to examine the potential macroeconomic effects of Brexit. The GFPM results indicate that, with Brexit and depending on the scenario, the consumption of sawnwood in Britain would be 1.0–2.1% lower by 2030, 2.9–6.1% lower for wood-based panels, and 1.9–4.1% lower for paper and paperboard. With Brexit, the UK's net trade deficit in sawnwood decreases by 4.8–9.9% by 2030, 4.4–9.1% for wood-based panels, and 5.5–10.8% for paper and paperboard. The effects on industrial roundwood consumption and production within the UK are negligible. The consequences of Brexit are mostly within Europe and driven predominantly by reduced consumption within the UK itself. While the Brexit effects are greater under the pessimistic scenario, the overall effect on the global wood products industry is small, and it has little to no effect on global prices.

In these examples of GFPM applications to international trade, trade flows are modeled such that bilateral flows of products are not tracked, which is in line with the model set up in the standard database ('World' file) of the GFPM. However, the GFPM can also be set up to quantify bilateral trade flows. This is done by disaggregating the import and export data in the World file by country of origin and destination. For example, Turner *et al.* (2001) modified the GFPM World file to allow multilateral trade flows among the countries of AFTA–CER and P5 to evaluate the effect on the New Zealand forest sector of tariff elimination under a trade agreement among these countries. AFTA–CER refers to the Association of South East Asian Nation (ASEAN) Free Trade Area – Closer Economic Relations; the CER countries are Australia and New Zealand. The P5 countries are United States, Chile, Australia, New Zealand, and Singapore. Turner *et al.* (2001) also studied the effect on New Zealand's forest sector of tariff reductions under the World Trade Organization (WTO) administered General Agreement on Tariffs and Trade (GATT).

The GFPM has also been used to study the effects of non-tariff barriers, such as quotas (see Chapter 5), embargoes, sanctions, and export bans. For example, Turner *et al.* (2008a) assessed the effects of removing non-tariff barriers on New Zealand exports of secondary processed wood products to the United States, China, and Japan. The calculated increase in trade value from improved market access is a modest 0.0%–9.2% of New Zealand's total wood products trade by 2030. Li *et al.* (2007) used the GFPM to project the impact on the world's forest sector of curtailing roundwood trade to reduce the spread of exotic pests. With a ban on roundwood trade over 5 years, world consumer expenditures for wood products rise by 2.2% and world value added is unchanged. However, producer revenues decrease by 16% for Russia and by 10% for New Zealand. Value added decreases by 13% in Japan, 7% in Korea, and 4% in China.

6.7.3 Application in studies related to climate change issues

Key questions related to climate change and forestry that confront researchers and policymakers include the following

1. What is the potential of forests and forest industries to offset atmospheric greenhouse gases?
2. How do forest-based climate mitigation strategies (e.g. carbon offset payment leading to reduced or delayed harvests) affect the forest sector?
3. What are the effects of climate change on forest growth and productivity (e.g. due to longer growing seasons, increased CO_2 fertilization, and temperature- and precipitation-related stresses) and the attendant impacts on forest product markets?

The GFPM has been applied to investigate parts of those questions in conjunction with carbon estimating models/equations that utilize GFPM projected outputs of forest stock, forest products, consumption, trade, and prices, and with exogenous projections about climate change effects on forest productivity that serve as inputs to the GFPM in adjusting forest growth and stock and shifting timber supply. For instance, utilizing GFPM projections, Johnston *et al.* (2019) estimated the forest sector carbon mitigation potential of 180 countries during 2015–2065. The projected changes in forest carbon storage are driven by the dynamic relationships between endogenously determined roundwood harvests, forest growth and inventory, forest area, wood product consumption (domestic production plus imports minus exports), and exogenous demographic and income changes consistent with the IPPC-SPP2 scenario resembling recent historical trends. The results suggest that, while the forest sector was a carbon source in the past, it would achieve status as a net carbon sink by 2030.

Johnston and Radeloff (2019) also used GFPM to make projections, conditional on the IPCC socioeconomic pathways SSP1 to SSP5, to estimate the mitigation potential of carbon stored in harvested wood products (HWPs), including sawnwood, panels, paper, and paperboard. They found that the global HWP pool was a net annual sink of 335 Mt of CO_2 in 2015, increasing by as much as 441 Mt CO_2 per year by 2030. However, even under favorable socioeconomic conditions, carbon stored annually in HWPs is < 1% of global emissions. Furthermore, carbon stored within end-use HWPs varies widely across countries and depends on evolving market conditions.

In a similar study, Nepal *et al.* (2013a) applied the USFPM/GFPM to project changes in carbon sequestered in tree biomass and in wood products harvested from U.S. timberland. The projections are based on future forest stocks and forest products production, consumption, and trade in the U.S. under four future scenarios, including wood energy consumption, based on the IPCC's 4th Assessment and the 2010 USDA Forest Service RPA scenarios (A1B, A2, B2, and HFW). The method also demonstrates how the USFPM/GFPM market modeling system can be used to evaluate the impacts on U.S. forest carbon and forest products markets of hypothetical forest carbon offset policies, by allowing carbon offset payments to compete with traditional forest product prices (e.g. see Nepal *et al.*, 2013b,c).

Applying the same principles to global markets, Buongiorno and Zhu (2013) used the stand-alone GFPM to investigate the consequences of carbon offset payments to forestland owners. They found that offset payments of \$15–\$50 per tCO_2 applied in all countries increased CO_2 sequestration in global forests by 5–14 billion tons of CO_2 from 2009 to 2030. Limiting implementation to developed countries, environmental damage is exported from north to south, as developing countries harvest more and thereby decrease their store of CO_2. Substantially more CO_2 is sequestered by allocating a given budget to all countries rather than to developed countries only. Because offset payments increase wood prices relatively more than they decrease production, the timber revenues of forestland owners generally increase. In the few countries with timber revenues losses. they are more than compensated for by the offset payments.

In another application of the GFPM to current environmental issues, Buongiorno and Zhu (2015b) explored the long-term, *ceteris paribus* effects of potential CO_2 fertilization on the global forest sector. As input, they used the findings of Norby *et al.* (2005) about forest response to elevated CO_2. Accordingly, forest productivity is raised in the GFPM in proportion to the rising CO_2 in three alternative IPCC scenarios. Projections of forest area, forest stock and production, consumption, prices, and trade of products are projected with the GFPM for each scenario, with and without CO_2 fertilization. The main effect of CO_2 fertilization was to raise the level of the world forest stock in 2065 by 9%–10% for scenarios A2 and B2, and by 20% for scenario A1B. The rise in forest stock induced by fertilization was in part counteracted by its stimulation of the wood supply, which resulted in lower wood prices and increased harvests.

A subsequent paper by Buongiorno (2016) explored the potential long-term effects of climate warming on the global wood sector. The study used Way and Oren's (2010) synthesis, indicating positive responses of tree growth to higher temperature in boreal and temperate climates, and negative responses in the tropics. Changes in forest productivity were introduced in the GFPM, using Way and Oren's equations in accordance with the rising temperatures projected in three IPCC scenarios. Projections of forest stocks, production, prices, trade, and value added in industries were obtained with the GFPM for each scenario, with and without temperature changes from 2012 to 2065. In the three scenarios, the projected total world growing stock of forests in 2065 was hardly changed by the rise in temperature. However, the forest stock was 2%–6% higher in developed countries, while it was 3%–4% lower in developing countries. There were significant attendant changes in wood production, prices, trade, and value added in forest industries benefiting developed countries and harming the developing countries.

The GFPM approach to evaluate forest sector impacts of climate change leading to changes in productivity is also used currently to project such effects in the UNECE countries, based on the projected changes in net primary productivity in different regions under the scenario of unconstrained CO_2 emissions.

6.7.4 Application in studies related to other policy questions

Past and current versions of the GFPM have been used extensively to assess the consequences of many other policy issues regarding international forestry, forest industries, and international trade. For example, Zhu and Buongiorno (2002) applied the GFPM to project the effect of paper recycling in the United States on the international forest sector. They found that, while the United States and other major consumer countries gained in welfare, Canada and the main European producers lost. Buongiorno and Zhu (2014) and Nepal *et al.* (2019b) studied the role of planted forests in the global forest economy. Li *et al.* (2008) and Turner *et al.* (2008b) dealt with the long-term effects of eliminating illegal logging on the world forests, forest industries, and trade. Zhang *et al.* (2012) explored the domestic and foreign consequences of China's land tenure reform on collective forests. Prestemon *et al.* (2006, 2008) derived the implications for timber product markets and trade of invasive species, such as the Asian *Lymantria* in the United States.

Several studies used the GFPM in conjunction with other models to understand the carbon offset potential in the U.S. of climate mitigation strategies, such as increased use of lumber in non-residential construction (Nepal *et al.*, 2016), expanded wood energy use (Nepal *et al.*, 2015), and long-term timber set asides (Nepal *et al.*, 2013c). The GFPM has also been integrated with U.S. national and regional forest sector models to represent the competition for wood between traditional products and energy (e.g. see Nepal *et al.*, 2019c, Stokes *et al.*, 2016).

6.8 Summary and Conclusions

The purpose of this chapter was to provide a general introduction, several examples of applications, and further references about the historical development and use of GFPM. Interested readers can build upon this basis for future studies to help answer different policy questions relevant for the global, regional, and/or national forest sectors. As with the REPA trade model, one attractive aspect of the GFPM is that the software, data, and documentation are available free of charge for research. Thus, with both the REPA model and GFPM, all studies are, in principle, replicable; the basic data sets can be expanded and improved, and shared among different researchers, knowing that the method, with its strength and blemishes, is the same.

To some extent, the first part of this chapter is a short version of the full documentation (Buongiorno and Zhu, 2018a,b) of the GFPM, describing, after a brief history, the general structure, mathematical formulations, needed input data, and the techniques to calibrate and validate the model. It must be emphasized that considerable work can be and has been done with the standard data set ('World' file) available with the software and, in that case, there is no need to redo the calibration and validation. Regardless, some readers may be more interested in the representative examples of GFPM applications, with its standard data set, or as expanded by other researchers. The hope is that the examples adequately document the usefulness of the model as a policy analysis tool to encourage at least some readers to try the GFPM for their own purpose.

Four groups of representative studies have been presented to illustrate: (1) past and ongoing GFPM's application in global and national forest sector outlook studies; (2) evaluations of the consequences of tariff and non-tariff barriers in the international trade of forest products; (3) projection of the impacts of climate change and forest-based climate change mitigation strategies on forests and forest industries; and (4) miscellaneous studies dealing with other important questions such as the effects of an increase in global planted forest area, illegal harvests, and invasive species.

In evaluating these and future studies, one must keep in mind that the GFPM, like any other economic model, is only a rough representation of reality. Buongiorno *et al.* (2003) documented how historical replications, let alone future projections, can differ substantially from observation. This is especially true for international trade, which is magnified by errors in domestic supply and demand. A study of the sources of uncertainty done with a stochastic version of the GFPM indicated that parameter uncertainty was more important than errors in the initial data. Among the parameters, the errors in the supply and demand elasticities tended to dominate those of the forest area and stock equations, the input–output coefficients and manufacturing costs, and the trade inertia parameters (Buongiorno and Johnston, 2018b).

The GFPM has been criticized for too much aggregation in describing the forest resources, such as a lack of a differentiation of forest stock by species, a lack of forest age class distribution, and a lack of competition for land use between alternative uses (e.g. agricultural vs forest use). Yet, in the standard GFPM version, the endogenous rate of change of forest area follows a Kuznets environmental curve, which does simulate changes in land use on various economic development paths. Furthermore, the GFPM allows for exogenous changes in forest area, in conjunction with or instead of the endogenous Kuznets curve-dependent change. Such exogenous changes may reflect land use competition (obtained possibly from other models) that can be part of the GFPM data. This approach was followed in parts of the USDA Forest Service's 2010 RPA Assessment studies, where the USFPM/GFPM was run with the exogenously projected U.S. forest area that took into account competition for land among different uses. The GFPM software also allows for disaggregation of the forest stock by softwood and hardwood groups, as demonstrated by Ince *et al.* (2011) for the U.S. and by Schier *et al.* (2018) for all other countries.

Similarly, on the demand side, the stand-alone GFPM provides a less detailed representation of end products demand than would be desirable for some studies (e.g. the standard GFPM aggregates softwood and hardwood lumber). However, the GFPM software gives the flexibility to introduce more detailed commodities and/or products (Buongiorno and Zhu, 2018a). For instance, more detailed wood products have been added in the GFPM for the U.S. (Ince *et al.*, 2011) and for all 180 countries represented in the GFPM (Schier *et al.*, 2018).

Another difficulty lies in the estimation of the GFPM parameters. As shown above, the calibration procedure leads to input–output coefficients and manufacturing costs that differ by country. However, for lack of more data, the standard GFPM uses the same elasticities in all countries for the demand for end products (sawnwood, panels, paper, and paperboard) and the same elasticities for the supply of raw materials (wood, waste paper, other fibers). Only the intercept terms of the supply equations are calibrated for each country based on the observed base year consumption, production, and price data to ensure that the model solution replicates the base year data. However, recent work has been done to obtain country-specific demand elasticities to better fit historical data, while staying within the 95% confidence of elasticities obtained by pooling data across countries (Buongiorno, 2019). These or other country-specific, econometrically estimated demand or supply elasticities can be readily incorporated in future GFPM versions. Despite its limitations, the GFPM remains a useful tool for outlook studies and policy analysis of the forest sector, available completely and freely to all interested researchers for a wide range of applications in its standard form or with user-specific modifications.

References

Buongiorno, J. (1996) Forest sector modelling: a synthesis of econometrics, mathematical programming, and system dynamics methods. *International Journal of Forecasting* 12, 329–343.

Buongiorno, J. (2015) Global modelling to predict timber production and prices: the GFPM approach. *Forestry* 88, 291–303.

Buongiorno, J. (2016) Global forest sector modeling: application to some impacts of climate change. In: Baardsen, S., Eid, T. and Hoen, H.F. (eds) *Festschrift in Honor of Professors Ole Hofstad and Birger Solberg*. INA fagrapport 36. Norwegian University of Life Sciences, Ås, Norway.

Buongiorno, J. (2019) Country-specific demand elasticities for forest products: estimation method and consequences for long term projections. *Forest Policy and Economics* 106, 101967.

Buongiorno, J. and Johnston, C. (2018a) Potential effects of U.S. protectionism and trade wars on the global forest sector. *Forest Science* 64(2), 121–128.

Buongiorno, J. and Johnston, C. (2018b) Effects of parameter and data uncertainty on long-term projections in a model of the global forest sector. *Forest Policy and Economics* 93, 10–17.

Buongiorno, J. and Zhu. S. (2013) Consequences of carbon offset payments for the global forest sector. *Journal of Forest Economics* 19, 384–401.

Buongiorno, J. and Zhu. S. (2014) Assessing the impact of planted forests on the global forest economy. *New Zealand Journal of Forest Research* 44(Suppl. 1), S2.

Buongiorno, J. and Zhu. S. (2015a) Technical change in forest sector models: the Global Forest Products Model approach. *Scandinavian Journal of Forest Research* 30(1), 30–48.

Buongiorno, J. and Zhu. S. (2015b) Calibrating and updating the Global Forest Products Model (GFPM version 2015 with BPMPD). *Staff paper series #84*. Department of Forest and Wildlife Ecology, University of Wisconsin-Madison. Available at: http://labs.russell.wisc.edu/buongiorno/ (accessed January 19, 2016).

Buongiorno, J. and Zhu. S. (2017) Potential effects of a Trans-Pacific partnership on forest industries. *Forest Policy and Economics* 81, 97–10.

Buongiorno, J. and Zhu, S. (2018a) Calibrating and updating the Global Forest Products Model GFPM version 2017 (with BPMPD and base year 2015). *Staff paper series #89*. Department of Forest Ecology and Management, School of Natural Resources, College of Agricultural and Life Sciences, University of Wisconsin-Madison. Available at: http://labs.russell.wisc.edu/buongiorno/welcome/gfpm (accessed November 5, 2019).

Buongiorno, J. and Zhu, S. (2018b) Using the Global Forest Products Model GFPM version 2017 (with BPMPD and base year 2015). *Staff paper series* #88. Department of Forest Ecology and Management, School of Natural Resources, College of Agricultural and Life Sciences, University of Wisconsin-Madison. Available at: http://labs.russell.wisc.edu/buongiorno/welcome/gfpm (accessed November 5, 2019).

Buongiorno, J., Zhu, S., Zhang, D., Turner, J. and Tomberlin, D. (2003) *The Global Forest Products Model: Structure, Estimation and Applications*. Academic Press/Elsevier, San Diego, California.

Buongiorno, J., Raunikar, R. and Zhu, S. (2011) Consequences of increasing bioenergy demand on wood and forests: an application of the Global Forest Products Model. *Journal of Forest Economics* 17, 214–229.

Buongiorno, J., Rougieux, P., Barkaoui, A., Zhu, S. and Harou, P. (2014) Potential impact of a transatlantic trade and investment partnership on the global forest sector. *Journal of Forest Economics* 20, 252–266.

Buongiorno, J., Johnston, C. and Zhu, S. (2017) An assessment of gains and losses from international trade in the forest sector. *Forest Policy and Economics* 80, 209–217.

FAO (1997) FAO Provisional Outlook for Global Forest Products Consumption, Production, and Trade to 2010. Food and Agriculture Organization of the United Nations, Rome, Italy.

Gilless, K. and Buongiorno, J. (1987) PAPYRUS: a model of the North American pulp and paper industry. *Forest Science*, Monograph 28.

IIASA (2018) SSP Database. International Institute for Applied Systems Analysis. Available at: https://tntcat.iiasa.ac.at/SspDb (accessed June 3, 2018).

Ince, P.J. and Nepal, P. (2012) Effects on U.S. timber outlook of recent economic recession, collapse in housing construction, and wood energy trends. *USDA Forest Service General Technical Report FPL-GTR-219*. Forest Products Laboratory, Madison, Wisconsin.

Ince, P.J., Kramp, A.D., Skog, K.E., Spelter, H.N. and Wear, D.N. (2011) U.S. forest products module: a technical document supporting the forest service 2010 RPA assessment. *USDA Forest Service Research Paper FPL-RP-662*. Forest Products Laboratory, Madison, Wisconsin.

IPCC (2018) Scenario process for AR5. Available at: http://sedac.ipcc-data.org/ddc/ar5_scenario_process/index.html (accessed August 11, 2018).

Johnston, C.M.T. and Buongiorno, J. (2017) Impact of Brexit on the forest products industry of the United Kingdom and the rest of the world. *Forestry* 90, 47–57.

Johnston, C.M.T. and Radeloff, V.C. (2019) Global mitigation potential of carbon stored in harvested wood products. *Proceedings of the National Academy of Sciences of the United States of America* 116(29), 14526 –14531.

Johnston, C., Buongiorno, J., Nepal, P. and Prestemon, J. (2019) From source to sink: past changes and model projections of carbon sequestration in the global forest sector. *Journal of Forest Economics* 34, 47–72.

Kallio, M., Dykstra, D.P. and Binkley, C.S. (eds) (1987) *The Global Forest Sector: An Analytical Perspective.* John Wiley and Sons, New York.

Kim, J.B., Monier, E., Sohngen, B., Pitts, G.S., Drapek, R. *et al.* (2017) Assessing climate change impacts, benefits of mitigation, and uncertainties on major global forest regions under multiple socioeconomic and emissions scenarios. *Environmental Research Letters* 12, 045001.

Korhonen, J., Nepal, P., Prestemon, J.P. and Cubbage, F.W. (2020) Projecting global and regional outlooks for planted forests under the shared socio-economic pathways. *New Forests* (2020). doi: 10.1007/s11056-020-09789-z

Kuznets, S. (1955) Economic growth and income inequality. *The American Economic Review* 45, 1–28.
Latta, G.S., Sjølie, H.K. and Solberg, B. (2013) A review of recent developments and applications of partial equilibrium models of the forest sector. *Journal of Forest Economics* 19, 350–360.
Li, R., Buongiorno, J., Zhu, S., Turner, J.A. and Prestemon, J. (2007) Potential economic impact of limiting the international trade of timber as a phytosanitary measure. *International Forestry Review* 9(1), 514–525.
Li, R., Buongiorno, J., Turner, J.A., Zhu, S. and Prestemon, J. (2008) Long-term effects of eliminating illegal logging on the world forest industries, trade, and inventory. *Forest Policy and Economics* 10, 480–490.
Mészáros, C. (1999) The BPMPD interior point solver for convex quadratic problems. *Optimization Methods and Software* 11&12, 431–449.
Nakicenovic, N., Davidson, O., Davis, G., Grübler, A., Kram, T. *et al.* (2000) *Special Report on Emissions Scenarios: A Special Report of Working Group III of the Intergovernmental Panel on Climate Change.* Cambridge University Press, Cambridge, UK. Available at: http://www.grida.no/climate/ipcc/emission/index.htm (accessed November 19, 2019).
Nepal, P. and Prestemon, J.P. (2019) UNECE/FAO timber section forest sector outlook studies iii background paper: selected scenarios and preliminary results. Available at: http://www.unece.org/fileadmin/DAM/timber/meetings/2019/20190214/09 _Nepal_and_Prestemon_UNECE_Koli_02_06_2019_PN_JPP.pdf (accessed October 20, 2019).
Nepal, P., Ince, P.J., Skog, K.E. and Chang, S.J. (2013a) Projection of U.S. forest sector carbon sequestration under U.S. and global timber market and wood energy consumption scenarios, 2010–2060. *Biomass and Bioenergy* 45, 251–264.
Nepal, P., Ince, P.J., Skog, K.E. and Chang, S.J. (2013b) Projected U.S. timber and primary forest product market impacts of climate change mitigation through timber set-asides. *Canadian Journal of Forest Research* 43 (3), 245–255.
Nepal, P., Ince, P.J., Skog, K.E. and Chang, S.J. (2013c) Forest carbon benefits, costs and leakage effects of carbon reserve scenarios in the United States. *Journal of Forest Economics* 19(3), 286–306.
Nepal, P., Wear, D.N. and Skog, K.E. (2015) Net change in carbon emissions with increased wood energy use in the United States. *Global Change Biology Bioenergy* 7, 820–835.
Nepal, P., Skog, K.E., McKeever, D.B., Bergman, R.D., Abt, K.L. *et al.* (2016) Carbon mitigation impacts of increased softwood lumber and structural panel use for nonresidential construction in the United States. *Forest Products Journal* 66 (1), 77–87.
Nepal, P., Korhonen, J., Prestemon, J.P. and Cubbage, F.W. (2019a) Projecting global and regional forest area under the shared socioeconomic pathways using an updated environmental Kuznets curve model. *Forests* 10(5), 387.
Nepal, P., Korhonen, J., Prestemon, J.P. and Cubbage, F.W. (2019b) Projecting global planted forest area developments and the associated impacts on global forest product markets. *Journal of Environmental Management* 240, 421–430.
Nepal, P., Abt, K.L., Skog, K.E., Prestemon, J.P. and Abt, R.C. (2019c) Projected market competition for wood biomass between traditional products and energy: a simulated interaction of U.S. regional, national, and global forest product markets. *Forest Science* 65(1), 14–26.
Norby, R.J., DeLucia, E.H., Gielen, B., Calfapietra, C., Giardina, C.P. *et al.* (2005) Forest response to elevated CO_2 is conserved across a broad range of productivity. *Proceedings of the National Academy of Sciences of the United States of America* 102(50), 18052–18056.
Prestemon, J.P., Zhu, S., Turner, J.A., Bungiorno, J. and Li, R. (2006) Forest product trade impacts of an invasive species: modeling structure and intervention trade-offs. *Agricultural and Resource Economics Review* 35(1), 128–143.

Prestemon, J.P., Turner, J.A., Bungiorno, J., Zhu, S. and Li, R. (2008) Some timber product market and trade implications of an invasive defoliator: the case of Asian *Lymantria* in the United States. *Journal of Forestry* 106(8), 409–415.

Raunikar, R., Buongiorno, J., Turner, J.A. and Zhu, S. (2010) Global outlook for wood and forests with the bioenergy demand implied by Scenarios of the Intergovernmental Panel on Climate Change. *Forest Policy and Economics* 12, 48–56.

Riahi, K., van Vuuren, D.P., Kriegler, E., Edmonds, J., O'Neill, B.C. *et al.* (2017) The shared socioeconomic pathways and their energy, land use, and greenhouse gas emissions implications: an overview. *Global Environmental Change* 42 (2017), 153–168.

Schier, F, Morland, C., Janzen, N. and Weimar, H. (2018) Impacts of changing coniferous and non-coniferous wood supply on forest product markets: a German scenario case study. *European Journal of Forest Research* 137(3), 279–300.

Schrage, L. (1991) *Optimization Modeling with LINDO*, 5th edn. Duxburry Press, Pacific Grove, California.

Stokes, B., Rials, T.G., Johnson, L.R., Abt, K.L., Nepal, P. *et al.* (2016) At the roadside: forest resources. In: Langholtz, M.H., Stokes, B.J. and Eaton, L.M. (lead eds) *2016 Billion-Ton Report: Advancing Domestic Resources for a Thriving Bioeconomy, Volume 1: Economic Availability of Feedstock*. US Department of Energy, ORNL/TM-2016/160. Oak Ridge National Laboratory, Oak Ridge, Tennessee.

Takayama, T. and Judge, G. (1971) *Spatial and Temporal Price and Allocation Models*. North-Holland, Amsterdam, Netherlands.

Turner, J.A., Buongiorno, J., Horgan, G.P. and Maplesden, F.M. (2001) Liberalization of forest product trade and the New Zealand forest sector, 2000–2015: a global modelling approach. *New Zealand Journal of Forestry Science* 31(3), 320–338.

Turner, J.A., Buongiorno, J. and Zhu, S. (2006) An economic model of international wood supply, forest stock and forest area change. *Scandinavian Journal of Forest Research* 21, 73–86.

Turner, J.A., Buongiorno, J., Zhu, S. and Maplesden, F.M. (2008a) Effect of non-tariff barriers on secondary processed wood product trade: New Zealand exports to the United States, *China and Japan. New Zealand Journal of Forestry Science* 38 (2/3), 299–322.

Turner, J.A., Katz, A. and Buongiorno, J. (2008b) The economic implications of illegal logging for the New Zealand forest sector. *New Zealand Journal of Forestry* 53(2), 20–25.

USDA Forest Service (2012) Future of America's Forest and Rangelands: Forest Service 2010 Resources Planning Act Assessment. *General Technical Report WO-87*. U. S. Department of Agriculture, Forest Service, Washington, DC.

Way, D.A. and Oren, R. (2010) Differential responses to changes in growth temperature between trees from different functional groups and biomes: a review and synthesis of data. *Tree Physiology* 30, 669–688.

Zhang, D., Buongiorno, J. and Ince, P. (1996) A recursive linear-programming analysis of the future of the pulp and paper industry in the United States: changes in supplies and demands and the effects of recycling. *Annals of Operations Research* 68, 109–139.

Zhang, D., Buongiorno, J. and Ince, P.J. (1993) *PELPS III: a microcomputer price endogenous linear programming system for economic modelling*. Research Paper, USDA Forest Service. Forest Products Laboratory, Madison, Wisconsin.

Zhang, H., Buongiorno, J. and Zhu, S. (2012) Domestic and foreign consequences of China's land tenure reform on collective forests. *International Forestry Review* 14(3), 349–362.

Zhu, S. and Buongiorno, J. (2002) International impact of national environmental policies: the case of paper recycling in the United States. *International Forestry Review* 4(2), 133–142.

Zhu, S., Tomberlin, D. and Buongiorno, J. (1998) Global forest products consumption, production, trade, and prices: GFPM model projections to 2010. *Working paper GFPOS/WP/01*. Food and Agriculture Organization of the United Nations, Rome, Italy.

Zhu, S., Buongiorno, J. and Brooks, D.J. (2001) Effects of accelerated tariff liberalization on the forest products sector: a global modeling approach. *Forest Policy and Economics* 2, 57–78.

Zhu, S., Turner, J.A. and Buongiorno, J. (2008) Long-term economic impact of countervailing duties on coated free sheet paper imported by the United States from China, the Republic of Korea, *and Indonesia. Forest Products Journal* 58(10), 71–77.

7 Softwood Lumber Trade and Trade Restrictions: Gravity Model

Xintong Li, Fatemeh Mokhtarzadeh and G. Cornelis van Kooten

University of Victoria, Victoria, BC Canada

Abstract

A gravity trade model can be used to determine the effects of policy on bilateral trade flows. The gravity model is initially explained and then used to determine the effect that U.S. tariffs have on softwood lumber (SWL) imports from Canada, using information from the 2006 Softwood Lumber Agreement. Quarterly data for seven Canadian and three U.S. regions for the period 2007–2017 are used to estimate a gravity model of SWL trade. The model is subsequently expanded to include Japan and China as separate regions, and then as a combined China–Japan region. The model is estimated using OLS and a Poisson Pseudo-Maximum-Likelihood method for trade quantity and value. Findings indicate that: (1) the imposition of a countervailing and/or anti-dumping duty usually has a negative effect on Canada's physical exports, but not in all cases; (2) the value of softwood lumber trade decreases by 26% on average under a tax/tariff compared with no duties; (3) the tax/tariff has a smaller but still significant impact on Canadian exports when China and Japan are included, as SWL exports are diverted from the U.S.; and, not surprisingly, (4) duties affect the value of lumber exports to a much greater extent than quantity.

In 2017, global exports of forest products were valued at \$423.5 billion, with trade in industrial coniferous roundwood (softwood logs) and coniferous sawnwood (softwood lumber) valued at \$8.0 billion and \$27.5 billion, respectively (FAO, 2020). Global production of softwood logs amounted to about 717.8 million m^3, while that of softwood lumber reached 347.4 million m^3. Canada is a major producer and exporter of forest products, especially softwood lumber, accounting for an average of 16% of global production of logs and 14% of total lumber output over the period 2007–2017. In 2017, Canada accounted for some 22% of softwood lumber (SWL) exports but less than 1% of imports. Within Canada, British Columbia accounted for 54% of Canada's log production and 45% of softwood lumber production.

Lumber exports from Canada to the United States declined between 2006 and 2011 before beginning to recover, as indicated in Fig. 7.1. The decline in

DOI: 10.1079/9781789248234.0007

exports was primarily due to the collapse of the U.S. housing market and the financial crisis, although the 2006 Softwood Lumber Agreement (see Chapter 2, this volume) also negatively impacted exports to the U.S. Yet, the U.S. absorbed more than half of Canadian softwood lumber exports during those years, as the U.S. is Canada's largest SWL trading partner. SWL exports to Japan have been important, but the trend has been downwards, while exports to China increased significantly between 2006 and 2011 before levelling off (Fig. 7.1).

Canada's exports of softwood logs constitute less than one-fifth of softwood lumber exports, with log exports to the U.S., China, and Japan over the period 1997–2017 provided in Fig. 7.2. Canada's log exports to the U.S. and Japan have trended downwards since the early 2000s, while those to China rose dramatically after 2006, exceeding exports to Japan plus the U.S. after about 2010. As the volume of log exports remains small compared with lumber exports (van Kooten, 2014; van Kooten and Johnston, 2014), the U.S. continues to have the greatest impact on Canada's softwood lumber exports.

China is now a dominant player in the world's forest sector. It experienced rapid growth in forest products output (ca. 22% in 2011) and doubled its wood panel production in 4 years. It doubled pulp, paper, and paperboard production in the past decade and now accounts for 21% of global production. China has also become the largest producer of wood furniture in the world (accounting for 20% of global capacity), the second largest producer of wood plastic composites, and the second largest importer and exporter of wood products in the world after the United States (United Nations Economics Commission for Europe (UNECE), 2011). Yet, it is evident from Fig. 7.1 that the U.S. remains by far the most important market for Canadian softwood lumber.

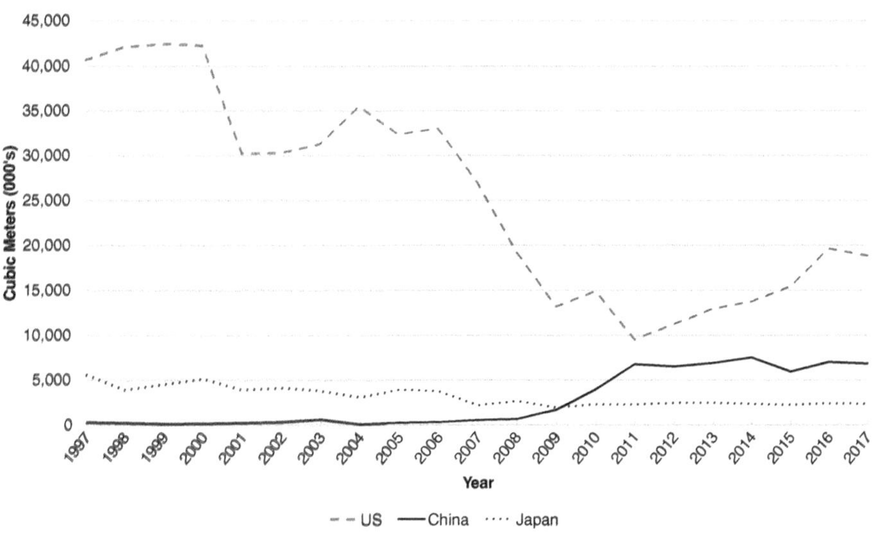

Fig. 7.1. Canadian softwood lumber exports to major markets, annual, 1997–2017. (Adapted from: Food and Agriculture Organization of United Nation, 2020.)

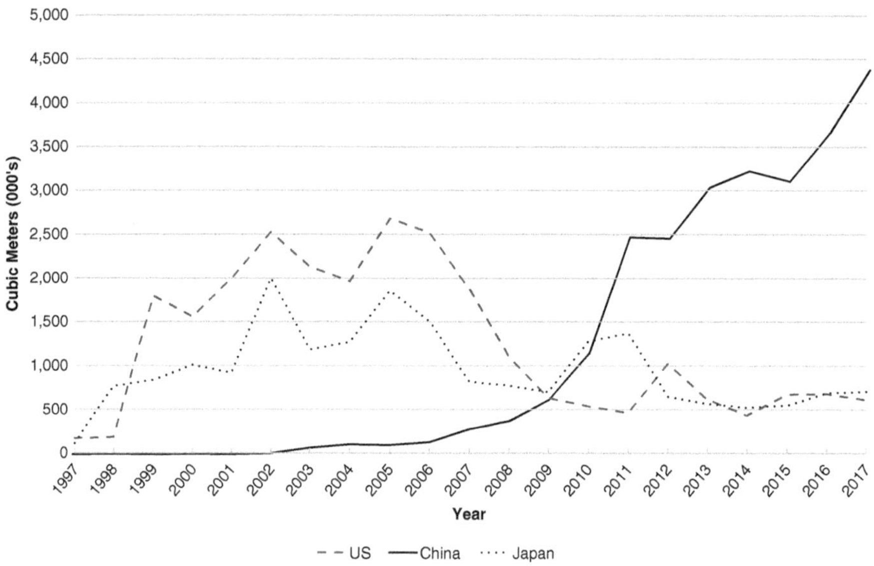

Fig. 7.2. Canadian industrial log exports to major markets, annual, 1997–2017. Adapted from: Food and Agriculture Organization of United Nation, 2020.)

Canadian shipments of lumber to the U.S. have historically been subject to dispute, with U.S. lumber producers claiming that the Canadian tenure system leads to implicit subsidies to lumber producers by providing them with low-cost logs. As noted in Chapter 2, the allocation of export quotas among provinces has been a source of controversy under the 2006 Softwood Lumber Agreement (SLA2006), because there were times when one or more regions failed to export their quota allotment, while another region exceeded its quota limit. Nonetheless, Canadian lumber exports to the U.S. declined after 2006 more as a result of the financial crisis than the quota system imposed under SLA2006.

Almost immediately upon expiration of SLA2006 in November 2016, the U.S. Lumber Coalition initiated trade action against Canada. As explained in Chapter 2, countervailing (CV) and anti-dumping (AD) duties were subsequently levied by the United States on softwood lumber from Canada, which led to complaints by the Canadian government under World Trade Organization (WTO) and the new United States–Mexico–Canada (USMCA) trade agreement. By the end of 2019, complaints were settled under WTO, but those under the new USMCA trade agreement are still underway.

The main purpose of the current chapter is not to focus on the complaint mechanisms and how the ongoing trade dispute might be resolved. Rather, the purpose is to develop a gravity model of softwood lumber (SWL) trade to estimate the effect that U.S. tariffs have on SWL trade between Canada and the U.S., focusing in particular on regional Canadian shipments of SWL to the U.S. The gravity trade model provides insights into the costs and benefits of the trade dispute, and how these might change under different combinations of CV and AD duties, even though the gravity equation cannot reflect the welfare loss directly.

In the next section, we briefly describe the basic concepts underlying the gravity model, explain the data collection process, and provide a summary of the major variables used to estimate the regression models. Then, in section 7.2, we develop the modeling approach – the gravity regression models – and discuss the ordinary least squares (OLS) and Poisson Pseudo-Maximum-Likelihood (PPML) methods that are subsequently used to estimate the model parameters. The data are described in section 7.3, with the estimation results provided in section 7.4; these provide insights regarding the application of OLS and PPML methods on different data sets. Sensitivity tests are also included in section 7.4. The final section discusses the limitations of our results and suggestions for further research.

7.1 Gravity Trade Model

In the international trade literature, gravity models are commonly used to quantify factors that influence trade. Similar to the gravity equations in physics, gravity models in economics include two major features – the regional gross domestic product (GDP) and the distances between regions, with the value of trade often also included as an important variable in the gravity equation. The first gravity model in economics was developed by Tinbergen in 1967 (Head and Mayer, 2014), with subsequent research finding that the gravity model fits trade data quite well. Anderson (1979, 2011) provided a theoretical foundation for using the gravity model for studying bilateral trade; subsequent to the 1979 paper, an influential structural gravity model was derived as a Ricardian structure by Eaton and Kortum (2002). Finally, the Armington-CES (constant elasticity of substitution) model of Anderson (1979) was popularized by Anderson and van Wincoop (2003).

The general gravity equation is given by:

$$X_{ij} = \alpha \frac{Y_i^{v_1} \times Y_j^{v_2}}{D_{ij}}, \tag{7.1}$$

where X_{ij} is the value of trade between regions i to j, with a positive value indicating export from i to j; Y_i is the GDP in region i; D_{ij} is the distance between regions i and j; and α, v_1, and v_2 are parameters to be estimated. In the traditional gravity model, one could expect that higher levels of GDP result in greater trade between countries or regions – economic growth in two regions will enhance the trade between them. However, as the distance between two regions increases, the cost of trading also rises, which reduces the passion for trade. A characteristic of the traditional gravity model is that trade is determined only by variables related to any two trading partners, without the influence of other regions engaged in trading.

Generally, the structural gravity approach based on the Armington-CES model assumes each country produces differentiated goods and exports them to the rest of the world. The structural gravity equation starts with the consumers and producers problem. On the demand side, the CES utility function is given by (Yotov *et al.*, 2016):

$$U(c_{ij}) = \left[\sum_i \alpha_i^{\frac{1-\sigma}{\sigma}} c_{ij}^{\frac{\sigma-1}{\sigma}} \right]^{\frac{\sigma}{\sigma-1}}, \tag{7.2}$$

where c_{ij} refers to the consumption goods provided by region i and consumed in region j; σ is the elasticity of substitution; and α_i represents the share parameter. The budget constraint is:

$$\sum_i p_{ij} c_{ij} = E_j, \tag{7.3}$$

where prices $p_{ij} = p_i\, t_{ij}$ refer to the price in foreign country j of the commodity produced by region i, which is determined by multiplying the price in the exporting country (p_i) by the transaction costs t_{ij} in exporting the commodity from region i to the export region j. E_j is the importer's total expenditure on the commodity. Solving the consumer optimization problem gives:

$$X_{ij} = \left(\frac{\alpha_i p_i t_{ij}}{P_j} \right)^{(1-\sigma)} E_j \tag{7.4}$$

where X_{ij} is the nominal value of trade between regions i and j, and P_j is the CES consumer price index with $P_j = \left[\sum_i (\alpha_i p_i t_{ij})^{1-\sigma} \right]^{\frac{1}{1-\sigma}}$.

In the next step, we use the market clearing condition where the nominal income for region i (Y_i) must equal the global expenditure on the good, assuming region i is the only producer in the world:

$$Y_i = \sum_j X_{ij}, \forall j. \tag{7.5}$$

After summing all X_{ij}, the market clearing condition (7.5) becomes:

$$Y_i = \sum_j \left(\frac{\alpha_i p_i t_{ij}}{P_j} \right)^{1-\sigma} E_j. \tag{7.6}$$

At an aggregate level, the world expenditure (the world nominal income Y) is equal to the sum of the values of all produced goods:

$$Y = \sum_i Y_i. \tag{7.7}$$

Therefore, the final form of the structural gravity model is:

$$X_{ij} = \frac{Y_i E_j}{Y} \left(\frac{t_{ij}}{\Pi_i P_j} \right)^{1-\sigma} \tag{7.8}$$

$$\Pi_i^{1-\sigma} = \sum_j \left(\frac{t_{ij}}{P_j} \right)^{1-\sigma} \frac{E_j}{Y} \tag{7.9}$$

$$P_j^{1-\sigma} = \sum_i \left(\frac{t_{ij}}{\Pi_i} \right)^{1-\sigma} \frac{Y_i}{Y} \tag{7.10}$$

Equations (7.9) and (7.10) are referred to as the inward and outward multilateral resistance, respectively. These equations measure the ease of market access for importers and exporters. Since the multilateral resistance equations include trade costs with all trade partners, the structural gravity model links all of the trade partners. Hence, trades between countries are not isolated as in the traditional model but correlated instead (Yotov *et al.*, 2016).

Measuring the multilateral resistance leads to a new problem. Although rooted in economic theory, it is difficult to measure the values in equations (7.9) and (7.10) in practice. For example, trade costs include not only the shipping and handling costs and any tariffs, but also non-tariff trade barriers (including ones related to language), issues of contiguity, etc. (Yotov *et al.*, 2016). It is not an easy task to account for all the variables that contribute to the transaction costs involved in trade; even if the variables can be identified, data on the variables may not be available.

Several methods have been developed for estimating multilateral resistances. One approach is to use a proxy known as the 'remoteness index'. Wei (1996), and Baier and Bergstrand (2009), used GDP and distance for multilateral resistances – that is, as proxies for the 'remoteness index'. Following Yotov *et al.* (2016), we apply the following remoteness index constructed in 2003 by Head and Mayer (2014):

$$lnREM\_EXP_{i,t} = ln\left(\sum_j Dist_{ij} / \left(\frac{E_{j,t}}{Y_t}\right)\right) \tag{7.11}$$

$$lnREM\_IMP_{j,t} = ln\left(\sum_i Dist_{ij} / \left(\frac{Y_{i,t}}{Y_t}\right)\right) \tag{7.12}$$

However, Anderson and Wincoop (2003) argued that the remoteness indexes are just analogue to the multilateral resistance with no theoretical basis.

In addition to the use of a remoteness index, a second method employs exporter/importer-time fixed effects. This method groups the exporters and importers according to when trade occurs, using the time index as a fixed effect (Yotov *et al.*, 2016). For instance, Alberta in 2013Q1 (1st quarter of 2013) can be considered an exporter-time fixed effect (FE), and the U.S. North in 2013Q1 as an importer-time FE. This FE method could absorb the unobserved effect from certain importers and exporters within a specific time. However, in our data for Canadian and U.S. regions, the exporter-time FE absorbs the effect of the tariff, because the data set does not provide enough variation to distinguish the influence of the tariff. In the above example, when Canada trades with the U.S. only, there are four exporter-time FEs for Alberta (one FE for each quarter) during 2013. In 2013, tariffs were imposed on Alberta in Q3 and Q4, but not in Q1 or Q2. The effect of the tariff can be determined as the linear combination of the exporter-time FE for Alberta. Hence, the tariff's influence cannot be separated from the exporter-time FE. Due to data restrictions, the effects will show up in trade with the U.S. only.

After introducing SWL trade data with third parties, the U.S. tariff changes with country pairs, and the FE method takes this into account. When we add China and Japan as other trade partners for Canadian softwood lumber, the exporter-time FE is no longer a linear combination of the tariff's effect, and the effect of the export tariff and multilateral resistance is quantified separately. Therefore, we can use the exporter-time FE. This method provides a consistent estimator only when the regression is measured in the multiplicative form, and thus OLS estimation is not applicable for the exporter/importer with time FEs. This method can be easily implemented on a computer, as researchers only need to create dummy variables by grouping time and trade regions. However, when observations have large time and regional scales, applying the method generates excessive variables that may influence the estimation results.

Further, double demeans, iterative structural estimation, and ratio-type estimation are all developed for approximating the multilateral resistance. Here we apply the remoteness index method for all observations, and the FE estimation for samples with the third party (see Tables 7.5 and 7.7 below). In the current application of the gravity model, the trade flow is characterized by SWL exports and their value. Moreover, as the gravity equation is in product form, it is common to take the logarithm of each side to linearize the regression equation. Thus, the gravity model is estimated in both multiplicative and logarithmic form. The estimation methods are discussed below.

7.2 Estimating the Effect of Tariffs: Modeling Approach

In our application, we divide North America into regions that differ from simple provinces or states. In particular, Canada and the U.S. are divided and aggregated into seven and three regions, respectively. Canadian regions consist of Alberta, Ontario, Quebec, Prairies (Manitoba and Saskatchewan), Maritimes (New Brunswick, Newfoundland, Nova Scotia, and Prince Edward Island), BC coast, and BC interior (Global Affairs Canada, 2020). The U.S. is divided into North, South, and West; the states included in each region are provided in Table 7.1. Three data sets are then developed. For the first data set, seven Canadian regions export only to the three U.S. regions. The second data set adds mainland China and Japan as third parties to the first data set to increase variation (five regions to which Canada exports), while the third data set combines China and Japan into a single China–Japan (CJ) region (giving four export regions).

We specify the traditional gravity model with the tariff as an independent variable (Yotov *et al.*, 2016):

$$lnX_{ij,t} = \beta_1 ln\,Y_{i,t} + \beta_2 ln\,Y_{j,t} + \beta_3 EXR_t + \beta_4 lnDist_{ij} + \delta\tau_{i,t} + u_{ij,t}. \tag{7.13}$$

In multiplicative form, the model is written as:

$$X_{ij,t} = \exp(\beta_1 ln\,Y_{i,t} + \beta_2 ln\,Y_{j,t} + \beta_3 EXR_t + \beta_4 lnDist_{ij} + \delta\tau_{i,t}) \times u_{ij,t}. \tag{7.14}$$

The dependent variable $X_{ij,t}$ is the quarterly quantity/value of softwood lumber that is exported by Canadian region *i* to U.S. region *j* during 2007–2017. Y_{kt} is the GDP in region *k*; EXR_t represents the exchange rate that converts Canadian

Table 7.1. U.S. regions classification. (Authors' own table.)

North	Connecticut	Delaware	Illinois	Indiana
	Iowa	Maine	Maryland	Massachusetts
	Michigan	Minnesota	Missouri	New Hampshire
	New Jersey	New York	Ohio	Pennsylvania
	Rhode Island	Vermont	West Virginia	Wisconsin
South	Alabama	Arkansas	Florida	Georgia
	Kentucky	Louisiana	Mississippi	North Carolina
	Oklahoma	Puerto Rico	South Carolina	Tennessee
	Texas	Virginia		
West	Alaska	Arizona	California	Colorado
	Idaho	Kansas	Montana	Nebraska
	Nevada	New Mexico	North Dakota	Oregon
	South Dakota	Utah	Washington	Wyoming

dollars to foreign currencies (the exchange rate is measured at the level instead of log form); and $Dist_{ij}$ represents the distances between export and import regions (distance is constant between any two regions but changes across pairs). The AD/CV tariff, $\tau_{i,t'}$ is a dummy variable that takes on a value of one when the tariff applies and zero otherwise. The coefficient δ determines the impact of the tariff.

Normally, the traditional gravity model includes other variables, such as continuity, common language, and colony. Because we apply the paired fixed effect in most of the models, the impact of these other variables is absorbed by the FE. Further, because the extra variables are not crucial factors, the traditional gravity model drops them from the estimation. Like the linear form, the traditional model in multiplicative form contains the same variables, but the dependent variable is in level instead of logarithmic form. Variables are included or eliminated to fit the specific models.

We begin by first estimating our models using OLS estimation for the additive form of the regression equations. The classic OLS estimator is used as a benchmark to compare other estimators. To estimate the regressions in multiplicative form, we adopt the PPML method. PPML was first applied to the gravity model by Silva and Tenreyro (2006), who argued that the traditional OLS estimator is biased and inconsistent when heteroskedasticity exists in the error term, even if the fixed effect is controlled. Compared with other estimators (OLS, Tobit, and nonlinear least squares), the PPML estimator leads to less bias in the Monte Carlo simulations. Second, Silva and Tenreyro (2006) found that the PPML method is able to deal with zero-trade observations, while with traditional OLS only non-zero trades are accepted, to avoid taking the logarithm of zero. Using the OLS estimator potentially imposes a positive constraint, making OLS estimation less desirable when the data include a large number of zero-trade values. With PPML, zero-trade observations can be preserved and unique information from the trade remains. Fortunately, in our first set of data (exports only to the U.S.), there are only two cases out of 924 observations where there is no trade. Omitting these two observations does not alter the main results of the OLS estimations. After adding Japan and China, the number of zero-trade samples increases and the PPML estimator is used instead.

As mentioned before, the paired FE is included in most of the estimations. This FE groups each exporter to each importer – for example, Alberta exports to the U.S. North constitutes one pair. In this way the unobserved time-invariant effect caused by trade-pairs is absorbed, which mitigates heteroskedasticity in error terms (Yotov *et al.*, 2016). Region-paired FE also addresses the endogeneity issue related to trade policy variables (Yotov *et al.*, p. 24). The data set where SWL trade only occurs between Canada and the U.S. contains 21 FE pairs. The data set that includes trade with China and Japan contains 35 FE pairs, while that which combines China and Japan into one CJ region has 28 FE pairs. Based on the paired FE features, distance variables are dropped in some estimations. We also extend the basic gravity model by adding time FEs in models – the time FEs model captures unobserved FEs and heteroscedasticity related to time. For the structural models, the time FE is omitted, because the approximation of multilateral resistances contains the 'time' information. The time FE consists of dummy variables representing 44 quarters (11 years). Similar to the paired FE regression, the exchange rate is removed when time FEs are present in the regression model.

In Tables 7.3 and 7.4 of the 'Estimation Results' (section 7.4) below, quantity is the dependent variable, while value is the dependent variable in Tables 7.5 and 7.6. The robust standard errors are clustered at the trade-pairs level in all of the models. Because provinces and states are gathered into various regions, it is necessary to take into account the correlation between trade-pairs. For instance, BC is divided into coastal (BCC) and interior (BCI) regions; thus, trade between BCC and other regions is highly correlated with exports from BCI. Clusters on trading pairs are used to take these possible correlations into account.

7.3 Data

The data used in this study come from different sources. We use quarterly data from 2007 to 2017 for each exporting region. The data on softwood lumber exports are extracted from Statistics Canada (2019a), with the Harmonized System code set at the six-digit commodity level (440710). Export prices are calculated by dividing export values by quantities. The prices and quantities are measured in Canadian dollars and in cubic meters, respectively. Table 7.2 provides data only for the case of Canadian exports to the U.S., while a description of the entire data that includes third parties is provided in Appendix Table 7.A.1.

Regional GDP data are mainly from Statistics Canada (2019b), the U.S. Bureau of Economic Analysis (2019), and the Organisation for Economic Co-operation and Development (OECD, 2019a,b). For the BC coast and BC interior specifically, the quarterly GDP data are estimated by multiplying total provincial GDP by the share of each region's population (BC Government, 2019a,b). The data on exchange rates are used to control for the volatility of

Table 7.2. Descriptive statistics. (Source: authors' calculations.)

Variable	Units	Obs	Mean	Std.Dev.	Min.	Max.
Quantity	'000s m^3	924	328	388	0	2,190
Price	C\$/m^3	924	156.24	40.87	0	300.61
Value	C\$×10^6	924	56.28	74.36	0	418.1
GDP Canada	C\$×10^9	924	256.21	203.18	35.00	825.81
GDP U.S.	US\$×10^9	924	5426.46	1281.55	3653.96	8275.76
Exchange rate	US\$/C\$	924	0.9016	0.0963	0.7225	1.037
Housing starts U.S.	units	924	77,371.2	34,519.1	26,323.6	187,000.0
Housing starts Can	units	924	7,070.3	5,660.6	128.5	23,248.0
Distances	km	924	2,282.3	1,102.6	294.6	4,255.1
Exchange rate	C\$/US\$	924	0.902	0.096	0.723	1.037
Tariff	=1 if in place	924	0.682	0.466	0	1

exchange rates since the output is measured in each regions' currency (OECD, 2019c,d,e). The U.S. population data are from the U.S. Census Bureau (https://www.census.gov). The tariff data are summarized from the Government of Canada (2016).

Distances between Canadian and U.S. regions are derived from Google maps (Google, 2020). Since regions are quite large, it is necessary to find regional centroids to determine distances between regions. For some regions, we use the city with the highest population as the centroid; for example, we choose Toronto and Montreal as the centroids for Ontario and Quebec, respectively. As for the two regions in British Columbia, we pick the Fraser Surrey Docks in the Greater Vancouver Regional District for the BCC, since this port handles the majority of lumber and log shipments (Canada's Pacific Gateway, 2019), and the city of Prince George for the BCI, as a large number of lumber companies are located in or near the city. Since Shanghai is China's largest economic center (Chen and Han, 2016) and has the largest port in the world (Keller *et al.*, 2013, p.337), we employ its port as the centroid for China. For Japan, we employ the port of Tokyo, which is Japan's leading port (Yang and Chen, 2015).

McCallum (1995) defined the combined regional centroid in terms of a population-weighted longitude and latitude. This approach is used for regions that consist of more than one jurisdiction, such as U.S. and Canadian regions that combine information from two or more states or provinces, or to combine Japan and China into a single region. Alternatively, rather than population, one could use trade value to weight the location of various centers across a region. The approach uses as an analogy the center of mass formula used in physics (Brown, 2007):

$$v_{cm} = \left(1/\sum_i m_i\right)\sum_i m_i\, v_i, \tag{7.15}$$

where m_i refers to the mass associated with vector v_i and v_{cm} is the mass-weighted aggregated vector representing the total of all the vectors. When applying this formula in our context, vectors represent the location of major cities in each region, while the value of SWL imports from Canada represents mass. To determine the centroid for region k (R_k), the center of mass formula is transformed into (Hedges, 2019):

$$R_k = \frac{\sum_{r=1}^{n_k}(longitude_r, latitude_r) \times v_r}{\sum_{q=1}^{n_k} v_r}, \forall k, \tag{7.16}$$

where n_k refers to the number of locations (states, provinces or countries) comprising region k that need to be aggregated, and $v_r = I_r/I_k$, with I_r representing a location r's (state r's) imports of SWL from Canada and I_k representing the SWL imports of the entire region k. We use the sum of SWL imports from Canada over the period 2007–2017.

To implement the above procedure for the three U.S. regions, we first need to determine state-level SWL imports. Within each of the regions identified in Table 7.1, we rank states according to the level of their SWL imports from Canada. Then, for the states in each region with the highest SWL imports and accounting for 70% or more of a region's SWL imports (usually requiring choosing five to seven states), we identify the cities with the highest population, using the geographical locations (longitude, latitude) of these cities as the main points to represent their states. The following example illustrates how the regional centroid is then determined.

Consider the U.S. West region as an example. Within this region, the shares of regional SWL imports (v_r) accruing to Washington, Oregon, Arizona, California, and Colorado were 35%, 13%, 9%, 8%, and 6%, respectively. These five states accounted for 72% of the U.S. West's total SWL imports. Seattle, Portland, Phoenix, Los Angles, and Denver were selected as the major cities for each of their respective states. Seattle is located at latitude 47.6° N and longitude 122.3° W, which is then multiplied by Washington's share of imports divided by the share of total regional SWL imports accounted for by the five states (72%); thus, the weight applied to Washington's location is 0.486 (= 0.35/0.72). The same is done for the other five top importing states, which enables us to identify the centroid for the U.S. West region as located at latitude 43.3° N, longitude 119.1° W. The same method was used to calculate centroids in the U.S. North and U.S. South regions, the Canadian Prairies and Maritimes, and for combining China and Japan into a single region.

One sensitivity test employs a different approach to calculating the distances between regions. Because U.S. regions contain more than five states, it is computationally difficult to employ the quarterly trade values rather than annual ones. Indeed, we employ trade values summed over the years for which we have data. In that case, distances between regions remain constant over time. In the third data set, however, only two regions (China and Japan) are aggregated. In this case, we can use quarterly import data from the seven Canadian regions to China and Japan to create a centroid that varies from one quarter to the next, because the import shares of Canadian SWL used to weight the latitude–longitude locations of the ports of Shanghai and Tokyo (namely v_C and v_J) change over time.

Besides distance, another sensitivity test replaces GDP with housing starts for trade to U.S. regions only. Although GDP is commonly used in simple gravity models, housing starts offer a different perspective on the trade relationship. Quarterly housing starts data are approximated by housing permits, with housing-related data collected from Statistics Canada (2019c) and the U.S. Census Bureau (2019a,b,c,d). Although correlated with GDP, housing starts are more directly related to the softwood lumber market than GDP, providing some notion of changes in buying patterns (e.g. switch from single to multiple family structures). An increase in house construction increases the demand for Canadian SWL, and vice versa. U.S. GDP, total and regional U.S. housing starts, and total Canadian exports of SWL to the U.S. are provided quarterly for the period 2007–2017 in Fig. 7.3. After 2008, regional housing starts in the U.S. exhibit a common upward trend, along with GDP and lumber imports from Canada. Housing starts do exhibit seasonal trends, however, with seasonally-adjusted housing starts tracking U.S. GDP quite closely.

In China, the number of wood structures has declined considerably, as concrete and steel have been preferred in light of rapid urbanization and industrialization, with the use of timber restrained after 1989 (Hu *et al.*, 2015). In 2009, only 500 wood-structured homes have been built, compared with more than

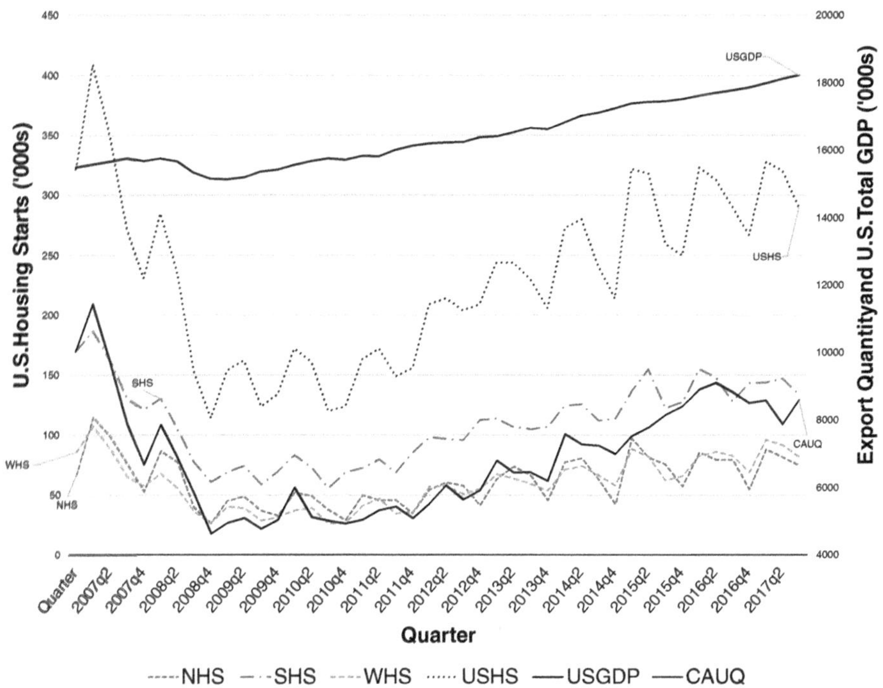

Fig. 7.3. U.S. Housing Starts, Total (USHS) and by Region (NHS, WHS, SHS), Canadian Softwood Lumber Exports (CAUQ), and U.S. GDP. (Adapted from: Statistics Canada 2019c; U.S. Bureau of Economic Analysis 2019; and U.S. Census Bureau 2019a,b,d.)

10 million apartments (Wang, 2009). In China, softwood lumber is not used mainly for construction, with the correlation between Chinese housing starts and softwood lumber imports much weaker than in the U.S. Therefore, we do not consider housing starts in regression models involving China and Japan.

7.4 Estimation Results

U.S. GDP clearly has a positive and significant influence on Canadian SWL exports, since the U.S. accounts for more than half of Canada's SWL exports (Fig. 7.1). But the effect of Canada's GDP on its SWL exports is not known a priori. On the one hand, postulating a positive relationship seems quite reasonable, because a larger economy implies a larger forest manufacturing sector and potential economies of scale that reduce prices. Contrarily, one could argue that the sign should be negative, because an increase in domestic GDP implies that more lumber is sold in the Canadian market, thereby increasing domestic prices and thus reducing exports to the U.S. Further, we expect that the exchange rate (US$/CAN$) is negatively correlated with exports. An increase in the exchange rate increases the number of Canadian dollars that American consumers get for each U.S. dollar, which effectively reduces the U.S. currency price of lumber. Finally, distance is an important factor in the gravity model as it roughly represents the transaction costs between trading partners. Naturally, an increase in the distance between two regions should reduce the volume of trade between them as transaction costs are higher. The relationship is obvious as one finds the BC Coast exports much more lumber to the U.S. West than to other regions, while Quebec sells more lumber in northeastern states and virtually nothing to the U.S. West.

All regression models are estimated using two dependent variables – export quantity and export value. The regression results are provided in Tables 7.3 to 7.6. It turns out that results are similar regardless of whether quantity or value is used as the dependent variable. The OLS results pertain to the traditional form of the gravity equation; the OLS regression equations are specified in models (1) through (4) in the Appendix, and models (5), (7) and (10) are also in the traditional form but estimated using PPML. The PPML estimates pertain to the structural gravity model (discussed above), with the PPML regression equations specified in models (6), (8), (9), (11) and (12) in the Appendix.

7.4.1 Trade quantity

When the dependent variable is export quantity, the basic OLS regression results associated with regression model (1) in Table 7.3 indicate that Canadian exports of softwood lumber to the U.S. are highly statistically sensitive to U.S. GDP (GDP_{imp}) and the US$/CAN$ exchange rate. An increase in U.S. GDP and a reduction in the exchange rate, which effectively reduces the price U.S.

Table 7.3. Quantity as dependent variable, Canadian exports to the U.S. only[a]. (Authors' own table.)

	OLS (Models 1 to 4)[c]				PPML (Models 5 and 6)[c]	
	(1)	(2)	(3)	(4)	(5)	(6)
Variable[b]	Basic	Paired FE	Time FE	Paired & Time FE	Paired & Time FE	Remoteness[d]
$\ln GDP_{imp}$	1.2483***	3.2527***	3.311**	0.5827	3.7811**	5.7849***
	(0.2966)	(0.9112)	(1.4784)	(2.4701)	(1.4887)	(1.1429)
$\ln GDP_{exp}$	−0.0597	−2.4599**	0.0599	−0.7583	−0.7334**	−5.6202***
	(0.233)	(0.9198)	(0.2318)	(0.9805)	(0.3735)	(1.341)
Exchange rate	−1.6706***	−0.9415***				−1.1345***
	(0.5392)	(0.2664)				(0.1902)
lnDist	−0.723		−0.4702			
	(0.4356)		(0.4429)			
τ (tax)	0.1866	−0.1744***	0.4302	−0.2534*	−0.0007	−0.0265
	(0.4104)	(0.0537)	(0.7508)	(0.1375)	(0.05)	(0.0228)
Observations	922	922	922	922	924	924
Pair dummy		YES		YES	YES	YES
Time dummy			YES	YES	YES	
Remoteness[d]						YES
Reset test (p value)	0.3058	0.2007	0.0059	0.0316	0.2607	0.0908

[a]The standard errors are provided in parentheses; statistical significance ***p<0.01, **p<0.05, *p<0.1.
[b]Explanatory variables where, e.g., $\ln GDP_{imp}$ refers to the logarithm of the lumber importer's GDP. GDP of importers (and exporters) will differ by region and year.
[c]Models refer to the regression equations found in the Appendix.
[d]Equations (7.11) and (7.12) are used to develop the remoteness index, which is the dependent variable in models 6 and 11 in the Appendix.

consumers pay for lumber, will increase exports of Canadian SWL to the U.S. However, the imposition of U.S. import tariffs or Canadian export taxes, both a consequence of the softwood lumber dispute, has no statistically discernible effect on Canadian lumber shipments to the U.S. This is somewhat surprising, although the effect changes when control variables are added to the model, as shown below.

The paired fixed-effects model (2) in Table 7.3 adds dummy variables to control for each pair of trading partners. In that case, the OLS estimation results indicate that the imposition of an export tax or import tariff lowers SWL exports from Canada to the U.S. by 16% ($e^{-0.1744}-1 = -0.16$) compared with no tax/tariff; this effect is significant at a 1% level. Then regression equation (3) in Table 7.3 captures the unobserved FE for each quarter, but the tax/tariff again has no statistically significant impact on Canada–U.S. lumber trade. In regression model (4), both the paired FE and the time FE are included, and the model is referred to as a two-way FE model. The effect of the tax/tariff now appears to have a larger impact on exports than in model (1), reducing them in this case by some 22% ($e^{-0.2534}-1=0.2238$). However, the estimated parameter (–0.2534) in model (4) is significant at the 10% level. Thus, both models (2) and (4) share statistically significant effects from tax/tariff and similar estimated parameters around –20%.

Regression equations (5) and (6) in Table 7.3 are estimated using PPML. In both models, the tax/tariff has no statistically significant effect on trade, but the GDPs of both the importing and exporting regions have a statistically significant impact on trade – a higher GDP in the importing countries increases trade, while a higher GDP in the exporting countries reduces trade. When remoteness is taken into account (model 6), the statistical significance of both GDP variables improves and the magnitude of the estimated coefficients is also greater. In this regression, the exchange rate turns out to be statistically significant as well, with trade declining with higher exchange rates.

Now consider Table 7.4 where the PPML method of estimation is employed in models (7) through (12). In this case, more observations are included: Canada trades SWL not only with the U.S. but also with one or two other jurisdictions. The first three regression equations, (7) through (9), in Table 7.4 include China

Table 7.4. Quantity as dependent variable, Canadian exports to the U.S., China and Japan[a]

	PPML (Models 7 to 12)					
	Exports to U.S., China & Japan			Exports to U.S. & China–Japan Region		
	(7)	(8)	(9)	(10)	(11)	(12)
Variable	Paired & Time FE	Remoteness	Exp-time or Imp-time FE[b]	Paired & Time FE	Remoteness	Exp-time or Imp-time FE[b]
$\ln GDP_{imp}$	1.3692***	1.3385***		2.2658***	1.8201***	
	(0.1129)	(0.1369)		(0.1639)	(0.168)	
$\ln GDP_{exp}$	–1.5129**	–0.673*		–1.4896**	–1.1911***	
	(0.7514)	(0.3483)		(0.6628)	(0.3141)	
Exchange rate	0.0028	0.0132***	0.0582***			
	(0.0037)	(0.0018)	(0.0027)			
τ (tax)	–0.0255	–0.0704**	–0.6149*	–0.0392	–0.0713**	–0.6901**
	(0.0523)	(0.0324)	(0.3331)	(0.0467)	(0.0298)	(0.3403)
Observations	1496	1496	1496	1232	1232	1232
Time dummy	YES			YES		
Pair dummy	YES	YES	YES	YES	YES	YES
Remoteness		YES			YES	
Exp-T or Imp-T[b]			YES			YES
Reset test (p value)	0.2710	0.0827	0.0190	0.3223	0.2130	0.0023

[a]See notes on Table 7.3.
[b]This refers to the exporter-time or importer-time fixed effect as discussed in Section 7.2.

and Japan as separate regions, while regression models (10) through (12) merge China and Japan into a single region (CJ). Again, the GDP of importing countries is positively correlated with increased trade, while the GDP of Canadian exporting regions tends to reduce trade. These impacts are statistically significant. The estimated parameters associated with the exchange rate between Canada and its trading partners have the wrong sign, suggesting that a rise in the exchange rate (more US$, Yen or Yuan needed to purchase a CAN$) will increase trade. This is contrary to expectation and, indeed, the estimates in models (8) and (9) are highly statistically significant (p-value < 1%). However, the parameters are very small, so that changes in the exchange rate have only a minor impact on softwood lumber trade.

An increase in export taxes or import tariffs reduces SWL exports in all of the regression models in Table 7.4, although it is not statistically significant in models (7) and (10). Except for models (9) and (12), the introduction of a tax/tariff decreases Canada's SWL exports to all regions from 2.6% to 7.1%. For models (9) and (12), exporter-time or importer-time fixed effects are taken into account, as these both measure multilateral resistance to trade; but only the tax/tariff is considered to be a regressor of interest. In both models, the tax/tariff is statistically significant (p-value < 1%), but the estimated parameter is larger than in any of the other models in Table 7.4. The tariff has larger influence on the trade to all regions than in other models, roughly –46% ($e^{-0.6149}-1 = -0.4593$) and –50% ($e^{-0.6901}-1 = -0.4985$) compared with –7.1%.

7.4.2 Value of trade

Next consider softwood lumber trade value as the dependent variable rather than quantity as the dependent variable. The regression results are provided in Tables 7.5 and 7.6, and they follow the same structure as Tables 7.3 and 7.4. First, we compare results in Table 7.5 with those in Table 7.3. The basic OLS models (model (1)) are similar. The values of estimated coefficients change slightly, but retain the same signs, while the influence of the tax/tariff to SWL export value (Table 7.5) drops to 9.0% from the 18.7% found in conjunction with export quantity (Table 7.3), although both are statistically insignificant. In model (3), the coefficient on the tax/tariff regressor has the wrong expected sign but it is not statistically significant as in model (1). The tariff's influence in models (2) and (4) is negative as expected and statistically significant (as in Table 7.3). From models (5) and (6), taxes/tariffs have a larger impact on trade value than trade quantity – respectively, –0.1% versus –4.9% (model (5)), and –2.7% versus –7.1% (model (6)). It is notable that the coefficient on the tariff is statistically significant at the 1% level in model (6). As in Table 7.3, PPML estimates of the tax/tariff influence are lower levels than with OLS.

From Table 7.5, we find that U.S. GDP has the correct sign (and is often highly statistically significant) across all six models, while the GDP of Canadian regions is negatively correlated with export value (but only statistically significant in three models), as expected. U.S. GDP appears to have had a greater

Table 7.5. Export value as dependent variable, Canadian exports to U.S. only[a]

	OLS (Models 1 to 4)				PPML (Models 5 and 6)	
	(1)	(2)	(3)	(4)	(5)	(6)
Variable	Basic	Paired FE	Time FE	Paired & Time FE	Paired & Time FE	Remoteness
$\ln GDP_{imp}$	1.6595***	5.256***	3.2837*	1.9768	4.6783***	6.81***
	(0.2886)	(0.8766)	(1.603)	(2.5533)	(1.2033)	(0.8682)
$\ln GDP_{exp}$	−0.0799	−2.6537***	0.0000	−0.7914	−0.7457**	−4.8965***
	(0.2292)	(0.8756)	(0.2462)	(0.9377)	(0.3341)	(1.0207)
Exchange Rate	−2.726***	−1.0372***				−1.3187***
	(0.5631)	(0.2702)				(0.1774)
lnDist	−0.7341		−0.5191			
	(0.4436)		(0.4708)			
τ (tax)	0.0902	−0.2352***	0.424	−0.2964**	−0.0494	−0.0712***
	(0.4126)	(0.0529)	(0.7695)	(0.1415)	(.0521)	(0.0181)
Observations	922	922	922	922	924	924
R-squared	0.990	0.999	0.992	0.999	0.980	0.924
Pair dummy		YES		YES	YES	YES
Time dummy			YES	YES	YES	NO
Reset test (p value)	0.335	0.204	0.009	0.033	0.080	0.047

[a]See notes on Table 7.3.

impact on trade values than on trade quantities. This former result is expected, since SWL export values directly relate to the region's expenditure levels. U.S. GDP has a greater impact than Canada's GDP, which is not unexpected.

Likewise, in Table 7.5, the signs on the exchange rate coefficients are negative as expected and statistically significant, with the exchange rate seemingly having a slightly higher influence on export values as opposed to export quantities, perhaps because the value regressions connect the market prices across two regions through trade.

Finally, Table 7.6 explains lumber exports from Canada to both U.S. regions and China and Japan; it can be compared with Table 7.4. As already noted, compared with using trade quantity, trade value as the dependent variable does not alter the major estimation results. The tax/tariff impacts trade value negatively in all models, although it appears from the higher coefficient estimates in Table 7.6 compared with Table 7.4 that trade value is influenced more by the tax/tariff than by trade quantity. What is more, the influence is now significant in models (7) and (10), while they are insignificant in Table 7.3. This means that imposing a tariff may not have direct influence on trade quantity but on SWL trade value instead. But this result is not unexpected, given that a tax/tariff directly affects value.

Overall, the estimated parameters in Table 7.6 are greater than those in Table 7.4, while the significant level stays similar. However, Canadian GDP

Table 7.6. Value as dependent variable, export to the U.S. and World[a]

	PPML (Models 7 to 12)					
	Canada Export to U.S., China, & Japan			Canada Export to U.S. & China–Japan		
	(7)	(8)	(9)	(10)	(11)	(12)
Variable	Paired & Time FE	Remoteness	Exp-time or Imp-time FE[b]	Paired & Time FE	Remoteness	Exp-time or Imp-time FE[b]
$\ln GDP_{imp}$	1.213***	1.1579***		1.455***	0.9057***	
	(0.0728)	(0.0553)		(0.1634)	(0.1296)	
$\ln GDP_{exp}$	–1.6285**	1.2518***		–1.7127*	1.3806***	
	(0.7906)	(0.209)		(0.8807)	(0.251)	
Exchange Rate	0.0037	0.0165***	0.1099***			
	(0.0029)	(0.0015)	(0.0023)			
τ (tax)	–0.0966**	–0.1316***	–0.6153**	–0.1332***	–0.1411***	–0.7127**
	(0.0395)	(0.0255)	(0.2701)	(0.0396)	(0.0281)	(0.2834)
Observations	1496	1496	1496	1232	1232	1232
Time Dummy	YES			YES		
Pair Dummy	YES	YES	YES	YES	YES	YES
Remoteness		YES			YES	
Exp-T (or Imp-T)			YES			YES
Reset test (p value)	0.2069	0.1784	0.0282	0.1133	0.0988	0.0009

[a]See notes on Table 7.3.
[b]See note on Table 7.4.

impacts trade value positively in models (8) and (11), while it reduces export quantity. Growth in the Canadian economy increases domestic demand for SWL, thereby raising lumber prices and, due to the substitution effect, the export quantity declines. When the effect from reduced quantity is fully compensated for by the increase in price, total trade value goes up.

Despite the foregoing result, whether the dependent variable is export quantity or value, the regression estimates display a similar pattern. When Canada trades only with the U.S., a tax/tariff has significant negative impact on trade value and trade quantity, around 20% for OLS and 4% for PPML and ranging from 0.07% to 29.64%. When trade with other jurisdictions is taken into account, the U.S. tariff still exhibits a significant influence on trade, which strengthens the importance of the tariff. The estimations matter for the results. Further, OLS estimates of the tax/tariff are higher on average than those using PPML. The exporter/importer-time FE offers comparably larger parameter estimates than the other models. Moreover, compared with estimating the impact on trade quantity, using the trade value as the dependent variable raises the absolute value of the tariff's influence. There is also some variation in the estimated coefficients on Canadian GDP, because trade value accounts for different information. Therefore, changing the dependent variable from quantity to value does not alter the tariff's impact on SWL trade.

7.4.3 Reliability tests

To check the reliability of models, several tests were conducted for each regression. The Ramsey RESET test is used to test for model misspecification; it is computed manually using the robust standard error, with p-values reported for each regression model at the bottom of the relevant tables. The null hypothesis states that the regression does not have a problem with omitted variables (Yotov *et al.*, 2016). Silva and Tenreyro (2006) found that PPML performed better in the Ramsey test than OLS. According to the results, two OLS models passed the test while three PPML models passed the test.

The second test was used to determine if there is an endogeneity problem, which occurs when a policy variable is related to the dependent variable. Recall that, under SLA2006, the export tax or import tariff was set to zero when the U.S. SWL price was high. As the export value of lumber is correlated with the market price, it is difficult to discover the effect of the tax/tariff, because of the feedback between market prices and the dependent variable, which is referred to as the 'reverse causality' between the tariff and trade. The endogenous trade policy variable would then bias the regression results. Since instrumental variables are not available, a test for 'strict exogeneity' is chosen to test for endogeneity. We do this by adding a fourth-quarter, forward-looking tariff (policy) variable to each model. When the estimated coefficient on the future level of the tariff equals zero from a statistical point of view, endogeneity with respect to the policy variable is not considered to be a problem (Yotov *et al.*, 2016). Results of this endogeneity test are provided in Table 7.7. With the exception of the paired and time FE models (7) and (10), and one remoteness index model (6), models fail to reject the null hypothesis that the coefficient on the added variable is zero. Hence, the tax/tariff does not, for the most part, indicate that there is a problem with endogeneity. Further, applying the paired FE also relieves the problem.

Other researchers have used regression analysis to analyze the U.S.–Canada SWL trade dispute by applying a variety of model structures. Parajuli *et al.* (2018) analyzed SLA2006 by using seemingly unrelated regression. Their research shows that the 2006 softwood lumber agreement negatively impacted the major Canadian lumber-producing provinces, reducing sales by some 4%–14% (at a 5% or 10% level of statistical significance), which coincides with the PPML results. Parajuli *et al.* (2015) estimated the overall impact applying a limited-information, maximum-likelihood method, concluding that the 2006 export tax did not have a statistically significant influence on Canadian SWL exports to the U.S.

Additional OLS estimation results are reported in the Appendix Tables 7.A.3. through 7.A.6 for softwood lumber trades that include exports to China and Japan, or China–Japan as one region. These help to reveal the entire story concerning Canadian softwood lumber exports.

Table 7.7. Test for endogeneity using the strict exogenous test[a]

		OLS (Models 1 to 4)				PPML (Models 5 and 6)	
		(1)	(2)	(3)	(4)	(5)	(6)
Model		Basic	Paired FE	Time FE	Paired & Time FE	Paired & Time FE	Remoteness
p-value	Quantity	0.409	0.589	0.522	0.277	0.914	0.001***
	Value	0.523	0.197	0.528	0.453	0.398	0.004***
		PPML (Models 7 to 12)					
		Canada Export to U.S., China & Japan			Canada Export to U.S. & China–Japan		
		(7)	(8)	(9)	(10)	(11)	(12)
Model		Paired & Time FE	Remoteness	Exp-time or Imp-time FE[b]	Paired & Time FE	Remoteness	Exp-time or Imp-time FE[b]
p-value	Quantity	0.054*	0.135	0.467	0.086*	0.107	0.391
	Value	0.022**	0.526	0.501	0.118	0.778	0.398

[a]Statistical significance indicated as: *** $p<0.01$, ** $p<0.05$, * $p<0.1$.
[b]See note on Table 7.4.

7.4.4 Sensitivity analysis

Sensitivity tests were conducted to demonstrate further the robustness of the results. These are provided in Tables 7.8 through 7.10. (In the tables, model *x.y* refers to model *x* from Tables 7.3 through 7.6, with *y* referring to the sensitivity version of *x*.) In Table 7.8, we consider transformations of model (4) from previous tables, including observations of trade from Canada to the U.S. only. The first two columns provide results from adding four lags of the dependent variable as regressors. Since data are quarterly, four lags are chosen to understand whether the previous year's SWL exports influence the current year's trade.

Sensitivity model (4.1) indicates that the tariff's negative influence drops from 22% (where $e^{-0.2534}-1 = -0.2238$) to 13% ($e^{-0.1425}-1 = -0.1328$) when a four-period lag is introduced; this is the least effect that the tax/tariff has among the OLS models. However, none of the variables of interest in model (4.1) are statistically significant. In model (4.2), which uses the trade value as the dependent variable, only the estimated coefficient on the tax/tariff variable is statistically significant, but only at the 10% level. The tax/tariff reduces exports by some 17%, somewhat less than the 22% reduction in the OLS regression with no lags of the dependent variable. The lagged export models suggest that the influence of tax/tariff may be slightly lower than indicated in models with no lags in the dependent variable. Nevertheless, adding lags to an FE model leads to an inconsistent and biased estimator (Baltagi, 2005), even if the sample has a long time length. Therefore, the lagged model is informative but imprecise.

Next, consider models (4.3) and (4.4) in Table 7.8. These replace the U.S. and Canadian GDPs with their respective housing starts. An increase in U.S.

Table 7.8. Sensitivity analysis: four-period lag on Canadian exports to the U.S. only[a]. (Authors' own table.)

	Lagged exports		Housing starts in lieu of GDP	
	Quantity	Value	Quantity	Value
	(4.1)	(4.2)	(4.3)	(4.4)
	Paired & Time FE	Paired & Time FE	Paired & Time FE	Paired & Time FE
$\ln GDP_{imp}$	–0.2841	0.4975		
	(1.3214)	(1.8224)		
$\ln GDP_{exp}$	0.1234	0.1138		
	(0.4012)	(0.4475)		
ln(Housing Starts$_{us}$)			–0.0411	0.0878
			(0.2032)	(0.2086)
ln(Housing Starts$_{ca}$)			–0.0275	–0.0346
			(0.0894)	(0.0794)
τ (tax)	–0.1425	–0.1852*	–0.2108*	–0.2524**
	(0.0895)	(0.1043)	(0.1138)	(0.1155)
Observations	838	838	922	922
Time dummy	YES	YES	YES	YES
Paired dummy	YES	YES	YES	YES
Four lags	YES	YES		

[a]Standard errors are in parentheses; statistical significance indicated as: ***p<0.01, **p<0.05, * p<0.1. Distance is included in the construction of the remoteness index.

housing starts should result in an increase in exports of SWL from Canada. In these two models, however, housing starts are statistically insignificant explanatory variables, with only the tax/tariff having a significant impact on Canadian SWL exports to the U.S., and then at the 10% and 5% levels of significance, respectively. The tax/tariff on trade is similar to what we found to be the case with models (1) and (2), explaining 19% ($e^{-0.2108}–1 = –0.1901$) and 22%($e^{-0.2524}–1 = –0.2231$) of the reduction in SWL exports, but smaller than the model (4) results reported in Tables 7.3 and 7.4. It appears that the use of regional GDP is more informative than using housing starts.

Tables 7.10 and 7.11 illustrate the sensitivity of the estimated results to a doubling of distances. Since we calculated the distances between two geographic points, it is possible that the true transaction distances are different from our approximation. Doubling the distances reflects how the results change with possible variations in distance. This test only applies when the effect of distances or remoteness indexes are measurable. Tests on doubled distances are from Anderson and Wincoop (2003). Upon comparing results in the two tables with the corresponding models in previous tables, it is clear that no considerable changes are caused by doubling the distances; even the parameter estimates change very little. Only in model (3.1) in Table 7.9 does the impact

Table 7.9. Sensitivity analysis: quantity as dependent variable, doubled distances. (Authors' own table.)

	Canada Export to U.S. only			U.S., China, & Japan	U.S., & China –Japan
	(1.1)	(3.1)	(6.1)	(8.1)	(11.1)
Variables	Basic	Time FE	Remoteness	Remoteness	Remoteness
$\ln GDP_{imp}$	1.2804***	3.3167**	5.7849***	1.3385***	1.8201***
	(0.3096)	(1.4314)	(1.1429)	(0.1369)	(0.1680)
$\ln GDP_{exp}$	–0.0589	0.0581	–5.6202***	–0.6730*	–1.1911***
	(0.2322)	(0.235)	(1.3410)	(0.3483)	(0.3141)
Exchange Rate	–1.6371***		–1.1345***	0.0132***	
	(0.5346)		(0.1902)	(0.0018)	
ln(2×Dist)	–0.7273	–0.0002			
	(0.4340)	(0.0001)			
τ (tax)	0.1867	0.4069	–0.0265	–0.0704**	–0.0713**
	(0.4089)	(0.6632)	(0.0228)	(0.0324)	(0.0298)
Observations	922	922	924	1,496	1,232
Time dummy		YES			
Pair dummy			YES	YES	YES
Remoteness			YES	YES	YES

[a]Standard errors are in parentheses; statistical significance indicated as: *** p<0.01, ** p<0.05, *p<0.1. Distance is included in the construction of the remoteness index.

of distance reduce the estimated coefficient dramatically, from around 0.47 in model (3) of Table 7.3 to –0.0002; however, in both cases the estimated coefficients are not statistically significant. Overall, both the OLS and PPML estimates are not sensitive to a doubling of distances.

We also constructed varied distances for the lumber trade with China and Japan, and China–Japan as a single region. The distances between each Canadian region and the China–Japan region change over time, because they contain the information of quarterly trade values from China and Japan. See Tables 7.A.7 and 7.A.8 in the Appendix. Compared with the fixed-distance models (10) through (12), the impact of a tax/tariff is increased, but still around the –10% for models (10) and (11) and –45% ($e^{-0.6000}-1 = -0.4511$) for model (12). After doubling their varied distances, the results remain similar to those found in the earlier Tables 7.3 through 7.6.

7.5 Discussion and Conclusions

Overall, in the fixed effects model, the imposition of countervailing and anti-dumping duties under SLA2006 had a statistically significant impact on softwood lumber export quantities and their value. Our results suggest that the value of SWL exports to the U.S. are on average 7%–30% lower than when there is no tax/tariff on Canadian exports. Since the tax/tariff variable is binary,

Table 7.10. Sensitivity analysis: value as dependent variable, doubled distances. (Authors' own table.)

	Canada export to U.S. only			U.S., China, & Japan	U.S., & China –Japan
	(1.1)	(3.1)	(6.1)	(8.1)	(11.1)
Variables	Basic	Time FE	Remoteness	Remoteness	Remoteness
$\ln GDP_{imp}$	1.6915***	3.3000**	6.8101***	1.1579***	0.9057***
	(0.3021)	(1.5594)	(0.8682)	(0.0553)	(0.1296)
$\ln GDP_{exp}$	–0.079	0.0013	–4.8965***	1.2518***	1.3806***
	(0.2287)	(0.2498)	(1.0207)	(0.2090)	(0.2510)
Exchange Rate	–2.6927***		–1.3187***	0.0165***	
	(0.5582)		(0.1774)	(0.0015)	
ln(2×Dist)	–0.7375	–0.0002			
	(0.4422)	(0.0002)			
τ (tax)	0.0905	0.4029	–0.0712***	–0.1316***	–0.1411***
	(0.4112)	(0.6793)	(0.0181)	(0.0255)	(0.0281)
Observations	922	922	924	1,496	1,232
Time dummy		YES			
Pair dummy			YES	YES	YES
Remoteness			YES	YES	YES

[a]Standard errors are in parentheses; statistical significance indicated as: *** p<0.01, ** p<0.05, *p<0.1. Distance is included in the construction of the remoteness index.

we are not able to examine the impact of different tax/tariff levels on Canadian lumber exports. Our findings suggest that, when additional export regions outside the U.S. are considered, specifically mainland China and Japan, the export tax/import tariff keeps lowering the softwood lumber export. Therefore, the effect of the tax/tariff remains statistically significant – the U.S. market remains the most important destination for Canadian softwood lumber.

Second, the estimated impact of a tax/tariff is affected by different statistical procedures. The maximum reduction in Canadian lumber exports to the U.S. is around 20% when OLS is used, but only 5% under the PPML method. Considering the scale of lumber exports to the U.S., policy makers cannot ignore the consequences of a 10% reduction in sales, given the importance of forestry to the Canadian economy.

Third, using trade quantity or trade value as dependent variables also influences results. Unsurprisingly, the tax/tariff has a greater effect on the value of trade than on the actual amount traded. When the trade quantity directly relates to Canadian softwood lumber production, the negative effect of the tariff impacts employment in the forest industry, which is the main concern of the Canadian government. Both trade quantity and value play crucial roles in the dispute.

Fourth, because the FE OLS model results in autocorrelation of residuals, lags of the dependent variable were added as regressors in an attempt to address this issue. This changed the tariff's impact slightly, compared with using the PPML estimator. However, the addition of lags in a fixed effect model is known to lead to biased results. Future researchers may wish to employ an alternative estimation method, such as the Arellano-Bond Estimator for the gravity model (see Baltagi, 2005; Wooldridge, 2010).

Finally, distances between exporting and importing regions were determined from the centroids of the various regions. Where regions consisted of sub-regions with data, we weighted the longitude and latitude of sub-regions by their share of total export values to derive the regional centroid. Distances were taken to be fixed over time, except for the China–Japan region where we permitted the centroid to vary quarterly according to the changes in the value of lumber imports from Canada for part of sensitivity tests. Recall that changes in the distances between exporting and importing regions had no statistically discernible effect on lumber movements. Future research could examine the use of state-level weighted distances. This would require identifying the major consuming municipality in each state, determining the quarterly value of imported lumber, and weighting these by their associated latitude and longitude. Even so, the distance variable is a proxy for shipping and handling costs with insignificant impact in the softwood lumber trade.

It is clear that this particular econometric approach to trade modeling gives one quite different insights into trade flows. The gravity model should be considered as a complement to the spatial price equilibrium models examined in the previous chapters. Researchers should be familiar with both approaches and, perhaps, yet another approach that has not been used in analyzing forest sector policies and their impacts on trade flows and regional welfares. This is the global vector autoregressive (GVAR) modeling approach, which is the topic of Chapter 8.

Appendix 7.A. Descriptive Statistics, Regression Models and Outputs

Descriptive statistics

Table 7.A.1. Descriptive statistics: exports to the U.S., China, & Japan. (Source: authors' calculations based on data collection.)

Variable	Unit	Obs	Mean	Std. Dev.	Min	Max
Quantity	'000s m^3	1,540	251.97	382.82	0	2,192.38
Price	C$/$m^3$	1,540	178.41	175.59	0	2,940.11
Value	C$×$10^6$	1,540	45.33	72.25	0	418.10
GDP Canada	C$×$10^9$	924	256.21	203.18	35.00	825.81
GDP China	US$×$10^9$	308	2,043.00	735.90	739.00	3,568.00
GDP Japan	US$×$10^9$	308	1,298.00	154.00	1,072.00	1,615.00
GDP China & Japan	US$×$10^9$	308	3,340.00	722.10	1,870,00	4,790.00
U.S. Dollar	USD/CAD	924	0.90	0.10	0.72	1.04
Chinese Yuan	CNY/CAD	308	5.96	0.74	4.69	7.54
Japanese Yen	JPY/CAD	308	89.98	10.54	73.76	115.88
Housing starts U.S.	units	924	77,371.20	34,519.10	26,323.60	187,000.00
Housing starts Canada	units	924	7,070.30	5,660.60	128.50	23,248.00
Distances	km	1,540	5,162.09	3,731.14	294.56	11,335.29
Tariff	=1 if tariff	1,848	0.34	0.47	0	1

Regression models

The following regression equations/models are reported in Tables 7.3 through 7.6 in the text.

Table 7.A.2. U.S. regions classification.

Notation	Variable	Remark	Notation	Variable
$X_{ij,t}$	Export quantity or value		θi	Trade paired effect
$Y_{i,t}$	Importer's GDP		γt	Time fixed effect
$Y_{j,t}$	Exporter's GDP		$REM\_exp_{i,t}$	Exporter's Remoteness effect
EXR_t	Exchange rate	$EXR_{ij,t}$ for the trade includes the third parties	$REM\_imp_{j,t}$	Importer's Remoteness effect
$Dist_{ij}$	Distances		$\pi_{i,t}$	Importer-time fixed effect
$\tau_{i,t}$	U.S. SWL tariff	$\tau i_{j,t}$ for the trade includes the third parties	$\chi_{j,t}$	Exporter-time fixed effect

OLS Specifications

Export to U.S. only

1. Basic

$$lnX_{ij,t} = \beta_1 ln\,Y_{i,t} + \beta_2 ln\,Y_{j,t} + \beta_3 EXR_t + \beta_4 lnDist_{ij} + \delta\tau_{i,t} + u_{ij,t} \quad \text{Model (1)}$$

2. Paired FE

$$lnX_{ij,t} = \beta_1 ln\,Y_{i,t} + \beta_2 \ln Y_{j,t} + \beta_3 EXR_t + \delta\tau_{i,t} + \theta_{ij} + u_{ij,t} \quad \text{Model (2)}$$

3. Time FE

$$lnX_{ij,t} = \beta_1 ln\,Y_{i,t} + \beta_2 ln\,Y_{j,t} + \beta_3 lnDist_{ij} + \delta\tau_{i,t} + \gamma_t + u_{ij,t} \quad \text{Model (3)}$$

4. Paired & Time FE

$$lnX_{ij,t} = \beta_1 ln\,Y_{i,t} + \beta_2 ln\,Y_{j,t} + \delta\tau_{i,t} + \theta_{ij} + \gamma_t + u_{ij,t} \quad \text{Model (4)}$$

PPML Specifications

5. Paired & Time FE:

$$X_{ij,t} = exp(\beta_1 ln\,Y_{i,t} + \beta_2 ln\,Y_{j,t} + \delta\tau_{i,t} + \theta_{ij} + \gamma_t) \times u_{ij,t} \quad \text{Model (5)}$$

6. Remoteness Index

$$X_{ij,t} = exp\,(\beta_1 ln\,Y_{i,t} + \beta_2 ln\,Y_{j,t} + \beta_3 EXR_t + \delta\tau_{i,t} + \beta_4 lnREM\_exp_{i,t} + \beta_5 lnREM\_imp_{j,t} + \theta_{ij}) \times u_{ij,t} \qquad \text{Model (6)}$$

Export to U.S., China & Japan

7. Paired & Time FE:

$$X_{ij,t} = exp(\beta_1 ln\,Y_{i,t} + \beta_2 ln\,Y_{j,t} + \beta_3 EXR_{ij,t} + \delta\tau_{ij,t} + \theta_{ij} + \gamma_t) \times u_{ij,t} \qquad \text{Model (7)}$$

8. Remoteness Index

$$X_{ij,t} = exp\,(\beta_1 ln\,Y_{i,t} + \beta_2 ln\,Y_{j,t} + \beta_3 EXR_{ij,t} + \delta\tau_{ij,t} + \beta_4 lnREM\_exp_{i,t} + \beta_5 lnREM\_imp_{j,t} + \theta_{ij}) \times u_{ij,t} \qquad \text{Model (8)}$$

9. Exporter-time (or Importer-time) FE

$$X_{ij,t} = exp\,(\beta_1\,EXR_{ij,t} + \delta\tau_{ij,t} + \pi_{i,t} + \chi_{j,t} + \theta_{ij}) \times u_{ij,t} \qquad \text{Model (9)}$$

Export to U.S. & China–Japan as one region:

10. Paired & Time FE:

$$X_{ij,t} = exp(\beta_1 ln\,Y_{i,t} + \beta_2 ln\,Y_{j,t} + \delta\tau_{ij,t} + \theta_{ij} + \gamma_t) \times u_{ij,t} \qquad \text{Model (10)}$$

11. Remoteness Index

$$X_{ij,t} = exp\,(\beta_1 ln\,Y_{i,t} + \beta_2 lnY_{j,t} + \delta\tau_{ij,t} + \beta_3 lnREM\_exp_{i,t} + \beta_4 lnREM\_imp_{j,t} + \theta_{ij}) \times u_{ij,t} \qquad \text{Model (11)}$$

12. Exporter-time (or Importer-time) FE

$$X_{ij,t} = exp(\delta\tau_{ij,t} + \pi_{i,t} + \chi_{j,t} + \theta_{ij}) \times u_{ij,t} \qquad \text{Model (12)}$$

Additional regression results

Table 7.A.3. OLS regression results with quantity as dependent variable, Canadian exports to the U.S., China, and Japan[a]

	(1)	(2)	(3)	(4)
Variable	Basic	Paired FE	Time FE	Paired & Time FE
lnGDPimp	–0.3799*	2.4031***	–0.2207	2.4196***
	(0.2046)	(0.6509)	(0.1679)	(0.6644)
lnGDPexp	1.1202***	–1.5636**	–0.5052	–3.3923**
	(0.3747)	(0.6301)	(0.6103)	(1.5058)
Exchange Rate	–0.0092	0.0099		–0.0005
	(0.0342)	(0.0078)		(0.0088)
lnDist	0.4021		–1.6603**	
	(0.5828)		(0.6864)	
τ (tax)	0.9555	–0.2067***	0.5671	–0.1674
	(0.7509)	(0.0685)	(0.7069)	(0.1453)
Observations	1540	1540	1540	1540
Pair dummy		YES		YES
Time dummy			YES	YES

[a]Statistical significance indicated as: *** p<0.01,** p<0.05,*p<0.1.

Table 7.A.4. OLS regression results with quantity as dependent variable, Canadian exports to the U.S. and the China–Japan region[a]

	(1)	(2)	(3)	(4)
Variable	Basic	Paired FE	Time FE	Paired & Time FE
$\ln GDP_{imp}$	–0.3132	2.6216***	–0.183	2.941***
	(0.1893)	(0.7116)	(0.1946)	(0.9043)
$\ln GDP_{exp}$	1.1708***	–1.7353*	–0.1699	–2.3861*
	(0.3344)	(0.9083)	(0.5726)	(1.2125)
lnDist	0.2385		–1.2125*	
	(0.5125)		(0.602)	
τ (tax)	0.5285	–0.2042***	0.5713	–0.1162
	(0.6548)	(0.0597)	(0.831)	(0.1474)
Observations	1232	1232	1232	1232
Pair dummy		YES		YES
Time dummy			YES	YES

[a]Statistical significance indicated as: ***p<0.01,**p<0.05,* p<0.1.

Table 7.A.5. OLS regression with value as dependent variable, Canadian exports to the U.S., China, and Japan [a]

	(1)	(2)	(3)	(4)
	Basic	Paired FE	Time FE	Paired & Time FE
$\ln GDP_{imp}$	–0.517*	3.5466***	–0.2998	3.5336***
	(0.2624)	(1.1835)	(0.2154)	(1.1989)
$\ln GDP_{exp}$	10.5965***	–1.2025	–0.3878	–4.8353**
	(0.5001)	(1.0187)	(0.8229)	(2.1818)
Exchange Rate	–0.0049	0.0207*		0.0059
	(0.0464)	(0.0113)		(0.012)
lnDist	0.5752		–1.9164**	
	(0.7288)		(0.8165)	
τ (tax)	0.9096	–0.2887***	0.4942	–0.2213
	(0.8979)	(0.092)	(0.7959)	(0.222)
Observations	1540	1540	1540	1540
Pair dummy		YES		YES
Time dummy			YES	YES

[a]Statistical significance indicated as: ***p<0.01, **p<0.05, *p<0.1.

Table 7.A.6. OLS regression results with value of exports as dependent variable, Canadian exports to the U.S. and the China–Japan region [a]

	(1)	(2)	(3)	(4)
Variable	Basic	Paired FE	Time FE	Paired & Time FE
$\ln GDP_{imp}$	–0.3986	2.8401***	–0.2437	2.6667**
	(0.2396)	(0.9991)	(0.2498)	(1.2001)
$\ln GDP_{exp}$	1.602***	–0.4211	0.0134	–2.9623*
	(0.4436)	(1.2017)	(0.7272)	(1.5422)
lnDist	0.3924		–1.2959*	
	(0.6067)		(0.677)	
τ (tax)	0.383	–0.2678***	0.5124	–0.0597
	(0.7473)	(0.0668)	(0.9167)	(0.2085)
Observations	1232	1232	1232	1232
Pair dummy		YES		YES
Time dummy			YES	YES

[a]Statistical significance indicated as: ***p<0.01, **p<0.05, *p<0.1.

Table 7.A.7. Doubled distances regression results with export quantity as the dependent variable, Canadian exports to the China–Japan region, varied distances [a]

	(1)	(2)	(3)	(4)	(5)	(6)
	Original	Doubled	Original	Doubled	Original	Doubled
Variable	Paired & Time FE	Paired & Time FE	Remoteness	Remoteness	Exp-time or Imp-time FE[b]	Exp-time or Imp-time FE[b]
$\ln GDP_{imp}$	0.1999	0.1999	0.5433	0.5433		
	(0.443)	(0.443)	(0.816)	(0.8161)		
$\ln GDP_{exp}$	−1.3941**	−1.3941**	−0.128	−0.128		
	(0.6485)	(0.6484)	(0.7225)	(0.7225)		
lnDist	13.7945***		8.0798***		11.4505***	
	(1.8944)		(2.8805)		(1.2167)	
ln(2×Dist)		13.7946***		8.0799***		11.4505***
		(1.8944)		(2.8805)		(1.2167)
τ (tax)	−0.1112**	−0.1112**	−0.0863***	−0.0863***	−0.603*	−0.603*
	(0.0542)	(0.0542)	(0.0331)	(0.0331)	(0.3357)	(0.3357)
Observations	1232	1232	1232	1232	1232	1232
Time dummy	YES	YES				
Pair dummy	YES	YES	YES	YES	YES	YES
Remoteness			YES	YES		

[a]Statistical significance indicated as: ***p<0.01,**p<0.05, *p<0.1.
[b]See note on Table 7.7.

Table 7.A.8. Doubled distances regression results with export value as dependent variable, Canadian exports to the China–Japan region, varied distances[a]

	(1)	(2)	(3)	(4)	(5)	(6)
	Original	Doubled	Original	Doubled	Original	Doubled
Variable	Paired & Time FE	Paired & Time FE	Remoteness	Remoteness	Exp-time or Imp-time FE[b]	Exp-time or Imp-time FE[b]
$\ln GDP_{imp}$	0.2413	0.2413	0.18	0.18		
	(0.4786)	(9.4786)	(0.7849)	(0.7849)		
$\ln GDP_{exp}$	−1.622*	−1.622*	1.9025***	1.9026***		
	(0.8626)	(0.8626)	(0.5276)	(0.5276)		
lnDist	8.1329***		2.2604		8.2713***	
	(2.1311)		(2.9809)		(1.0225)	
ln(2×Dist)		8.133***		2.2605		8.2714***
		(2.1311)		(2.9808)		(1.0225)
τ (tax)	−0.1773***	−0.1773***	−0.1465***	−0.1465***	−0.6473**	−0.6473**
	(0.0491)	(0.0491)	(0.0327)	(0.0327)	(0.2807)	(0.2807)
Observations	1232	1232	1232	1232	1232	1232
Time dummy	YES	YES				
Pair dummy	YES	YES	YES	YES	YES	YES
Remoteness			YES	YES		

[a]Statistical significance indicated as: ***p<0.01,** p<0.05,* p<0.1.
[b]See note on Table 7.4.

References

Anderson, J.E. (1979) A theoretical foundation for the gravity equation. *American Economic Review* 69(1), 106–116.

Anderson, J.E. (2011) The gravity model. *Annual Review of Economics* 3, 133–160.

Anderson, J.E. and van Wincoop, E. (2003) Gravity with gravitas: a solution to the border puzzle. *American Economic Review* 93(1), 170–192.

Baier, L. and Bergstrand, H. (2009) Estimating the effects of free trade agreements on international trade flows using matching econometrics. *Journal of International Economics* 77(1), 63–76.

Baltagi, B.H. (2005) *Econometric Analysis of Panel Data.* John Wiley & Sons Ltd., Chichester, UK.

BC Government (2019a) Building permits, housing starts & sales. Available at: https://www2.gov.bc.ca/gov/content/data/statistics/economy/building-permits-housing-starts-sales (accessed December 6, 2019).

BC Government (2019b) Municipalities, regional districts and development regions. Available at: https://www2.gov.bc.ca/gov/content/data/statistics/people-population-community/population/population-estimates (accessed December 6, 2019).

Brown, R.G. (2007) *Introduction of Physics I Elementary Mechanics.* Duke University Physics Department, Durham, North Carolina.

Canada's Pacific Gateway (2019) Forest products transportation infrastructure. Available at: https://www2.gov.bc.ca/gov/content/transportation/transportation-reports-and-reference/reports-studies/planning-strategic-economic (accessed July 20, 2020)

Chen, L. and Han, W. (2016) Shanghai: front-runner of community-based eldercare in China. *Journal of Aging and Social Policy* 28(4), 292–307.

Eaton, J. and Kortum, S. (2002) Technology, geography and trade. *Econometrica* 70(5), 1741–1779.

FAO (2020) FAOSTAT. Available at http://www.fao.org/faostat/en/#data (accessed July 20, 2020).

Global Affairs Canada (2020) Frequently Asked Questions – Softwood Lumber. Available at: https://www.international.gc.ca/controls-controles/softwood-bois_oeuvre/other-autres/faq.aspx?lang=eng (accessed July 20, 2020)

Google (2020) Distance between Canada and U.S. regional center. Available at: https://www.google.com/maps (accessed July 20, 2020).

Government of Canada (2016) Archived – Export Charge Rates for 2007. Available at: https://www.canada.ca/en/revenue-agency/services/tax/businesses/topics/softwood-lumber-products-export-charge/archived-export-charge-rates.html#hstrclrts (accessed December 6, 2019).

Head, K. and Mayer, T. (2014) Gravity equations: workhorse, toolkit and cookbook. *Handbook of International Economics* 4, 131–195.

Hedges, A. (2019) Convert latitude/longitude to decimal. Available at: https://andrew.hedges.name/experiments/convert_lat_long/ (accessed December 6, 2019).

Hu, Q., Dewancker, B., Zhang, T. and Wongbumru, T. (2015) Consumer attitudes towards timber frame houses in China. *Pocedia – Social and Behavioral Science* 216(6), 841–849.

Keller, W., Li, B. and Shiue, C. (2013) Shanghai's trade, China's growth: continuity, recovery, and change since the Opium Wars. *IMF Economic Review* 61(2), 363–378.

McCallum, J. (1995) National borders matter. *American Economic Review* 85(3), 615–623.

OECD (2019a) Current price gross domestic product in China (CHNGDPNQDSMEI). FRED, Federal Reserve Bank of St Louis. Available at: https://fred.stlouisfed.org/series/CHNGDPNQDSMEI (accessed December 6, 2019).

OECD (2019b) Current price gross domestic product in Japan (JPNGDPNQDSMEI). FRED, Federal Reserve Bank of St Louis. Available at: https://fred.stlouisfed.org/series/JPNGDPNQDSMEI (accessed December 6, 2019).

OECD (2019c) National currency to US dollar spot exchange rate for China (CCUSSP02CNQ650N). FRED, Federal Reserve Bank of St Louis. Available at: https://fred.stlouisfed.org/series/CCUSSP02CNQ650N (accessed December 6, 2019).

OECD (2019d) US dollar to national currency spot exchange rate for Canada [CCUSSP01CAQ650N]. FRED, Federal Reserve Bank of St. Louis. Available at: https://fred.stlouisfed.org/series/CCUSSP01CAQ650N (accessed December 6, 2019).

OECD (2019e) US dollar to national currency spot exchange rate for Japan (CCUSSP01JPQ650N). FRED, Federal Reserve Bank of St. Louis. Available at: https://fred.stlouisfed.org/series/CCUSSP01JPQ650N (accessed December 6, 2019).

Parajuli, R., Chang, S.J. and Hill, R.C. (2015) How effective is the United States–Canada softwood lumber agreement 2006? An econometric study. *Forest Science* 61(6), 1041–1049.

Parajuli, R., Zhang, D. and Kosman, K. (2018) Province specific impacts of the 2006 United States–Canada softwood lumber agreement: a seemingly unrelated regression approach. *Forest Policy and Economics* 90, 1–6.

Silva, J.S. and Tenreyro, S. (2006) The log of gravity. *The Review of Economics and Statistics* 88(6), 641–658.

Statistics Canada (2019a) *Table 980-0044 Domestic exports – wood and articles of wood; wood charcoal (table) CIMT (database).* Ottawa, Canada.

Statistics Canada (2019b) *Table 36-10-0104-01 Gross domestic product, expenditure based, Canada, quarterly (table) CANSIM (database).* Ottawa, Canada.

Statistics Canada (2019c) *Table 34-10-0135-01 Canada Mortgage and Housing Corporation, housing starts, under construction and completions, all areas, quarterly.* Ottawa, Canada.

UNECE (United Nations Economic Commission for Europe) (2011) Forest Annual Market Review 2010–2011. Available at: https://www.unece.org/fileadmin/DAM/publications/timber/FPAMR_2010-2011_HQ.pdf (accessed December 6, 2019).

U.S. Bureau of Economic Analysis (2019) SQGDP2 GDP by state 1/ GDP by state: all industry total. Available at: https://apps.bea.gov/itable/iTable.cfm?ReqID=70& step=1#reqid=70&-step=1&isuri=1 (accessed December 6, 2019).

U.S. Census Bureau (2019a) Building permits survey. Available at: https://www.census.gov/construction/bps/statemonthly.html (accessed December 6, 2019).

U.S. Census Bureau (2019b) Data relationships between permits, starts, and completions. Available at: https://www.census.gov/construction/nrc/nrcdatarelationships.html (accessed December 6, 2019).

U.S. Census Bureau (2019c) Definitions of division, new residential construction, and structure. Available at: https://www.census.gov/construction/nrc/definitions/index.html (accessed December 6, 2019).

U.S. Census Bureau (2019d) Started and completed by purpose of construction. Available at: https://www.census.gov/construction/nrc/historical_data/index.html (accessed December 6, 2019).

van Kooten, G.C. (2014) The benefits of impeding free trade: revisiting British Columbia's restrictions on log exports. *Journal of Forest Economics* 20(4), 333–347.

van Kooten, G.C. and Johnston, C.M.T. (2014) Global impacts of Russian log export restrictions and the Canada–U.S. lumber dispute: modelling trade in logs and lumber. *Forest Policy and Economics* 39, 54–66.

Wang, G. (2009) The present situation and prospects of timberwork houses in China. *Wood Processing Machinery* 2, 35–36.

Wei, S. (1996) Intra-national versus international trade: how stubborn are nations in global integration? *NBER Working Paper No. 5531*. National Bureau of Economic Research. Available at: https://www.nber.org/papers/w5531 (accessed December 6, 2019).

Wooldridge, J.M. (2010) *Econometric Analysis of Cross Section and Panel Data*. Massachusetts Institute of Technology, London.

Yang, Y. and Chen, S. (2015) Determinants of global logistics hub ports: comparison of the port development policies of Taiwan, Korea, and Japan. *Transport Policy* 45, 179–189.

Yotov, Y., Piermartini, R., Monteiro, J. and Larch, M. (2016) *An Advanced Guide to Trade Policy Analysis: The Structural Gravity Model*. United Nations and World Trade Organization, Geneva, Switzerland. Available at: https://www.wto.org/english/res_e/booksp_e/advanced-wtounctad2016_e.pdf (accessed November 7, 2019).

8 A Global Vector Autoregression Model for Softwood Lumber Trade

Fatemeh Mokhtarzadeh

University of Victoria, Victoria, BC Canada

Abstract

A novel econometric approach is developed in this chapter, namely, the Global Vector Autoregressive (GVAR) model. It provides a comprehensive framework for analyzing the country-level impacts of various domestic, foreign, and/or global shocks on softwood lumber trade. The GVAR approach is applied to Canada–U.S. trade in softwood lumber and used to analyze the effect of external shocks on Canadian lumber prices. Findings indicate that Canada's export prices are positively correlated to U.S. housing starts and real GDP. Further, using impulse response functions, it is used to examine the effects on regional lumber export prices in Canada of: (1) a change in U.S. housing starts; (2) a reduction in U.S. GDP by one standard deviation; (3) a COVID-19 induced decline in U.S. GDP (of three standard deviations); (4) an increase in global oil prices; and, in the Appendix, (5) an increase in the long-term interest rate. Price impacts vary a great deal by Canadian region depending on the type of shock, with the propagation mechanism in Alberta significantly different from that in other regions. For example, with an oil price shock and because Alberta is a major exporter of oil, the lumber export price remains high even as the shock dissipates over time.

Canadian exports of softwood lumber have stagnated over the past several decades, as indicated in Chapter 2, this volume (Figs 2.1 and 2.2). A wide range of explanations have been proposed for sluggish growth of softwood lumber exports, including lethargic global demand, a slowdown in housing starts, a shift from single-family to multiple-family housing, substitution of non-wood materials for wood in construction, widespread uncertainty about federal funding and mortgage rates in the U.S., and a failure to resolve the ongoing U.S.–Canada softwood lumber trade dispute. The importance of softwood lumber trade between Canada and the U.S. has been discussed in Chapters 1 and 2. In this chapter, the focus is on softwood lumber for two main reasons: first, it is the most traded wood product; and, second, many wood products are dependent on residues from lumber production, including wood chips for making pulp and oriented strand board (OSB), and fiber for medium density

10.1079/9781789248234.0008

board (MDF), other engineered products, and, of increasing importance, wood pellets to co-fire with or replace coal in generating electricity.

The literature has examined softwood lumber trade using two general approaches: the spatial price equilibrium (SPE) modeling approach, discussed in Chapters 4, 5 and 6; and an econometric method known as the gravity model (Chapter 7). The SPE approach to trade modeling seeks to maximize the total sum of consumer and producer surpluses, or, alternatively, minimize the total cost of trade. The SPE approach is discussed in Chapter 4; it employs a cross-sectional approach that is useful for quantifying the impact of various trade polices (primarily trade restrictions). However, it is not able to examine the effect of other factors on trade, such as housing starts, global shocks to oil prices, or the historical effect of changes in macroeconomic variables. To take these factors into account requires an econometrics approach that uses gravity models (Chapter 7) and statistical techniques such as ordinary least squares, vector autoregression, cointegration, seemingly unrelated methods, etc. Econometric approaches generally investigate the relationships among factors that might influence the volume of trade or the price of the product in question. When it comes to softwood lumber, factors such as real GDP, housing starts, and interest rates are generally of most importance.

The econometric approach enables us: (1) to study the relationship between variables in a time series fashion; and (2) to quantify the impulse responses to various shocks. In doing so, cointegration of various factors, short- and long-run relationships, and statistical significance of variables explaining softwood lumber trade are determined by employing different statistical tests (e.g. see Jennings *et al.*, 1991; Nagubadi and Zhang, 2013; Parajuli *et al.*, 2018; Babula and Zhang, 2019). For example, Nagubadi and Zhang (2013) studied U.S. imports of softwood lumber from Canadian provinces affected by American trade restrictions using cointegration methods. They found that, while U.S. domestic lumber prices and the housing market have a positive impact on lumber imports from affected provinces, higher Canadian lumber prices have a negative impact on U.S. imports from Canada in the long run. They also found that SLA1996 and SLA2006 (see Chapter 2) reduced U.S. imports of softwood lumber from affected Canadian provinces in the short run. These results are not surprising, however.

In a recent paper by Parajuli *et al.* (2018), the authors evaluated the effect of the SLA2006 on the U.S. lumber imports from Canadian provinces. They estimated U.S. demand for imports using seemingly unrelated regression at the provincial level. Their findings are indicative of the negative impact of SLA2006 on U.S. demand for softwood lumber from the affected provinces. Alternatively, Li *et al.* (Chapter 7, this volume) adopted a gravity model that used OLS and a Poisson Pseudo-Maximum-Likelihood method to quantify the effect of trade restrictions on Canadian provinces. Their model included seven Canadian regions and three U.S. regions, and, not surprisingly, they found that the tariff had a negative impact on Canada's export quantities and values.

When it comes to vector autoregressive (VAR) models, Jennings *et al.* (1991) studied the causal relationship between the Canadian lumber market and macroeconomic variables over the period 1968–1987. They found that housing starts have a significant effect on the level of lumber production in Canada, but that the exchange rate and interest rate had no statistically discernible effect. Their model is comprehensive in that they included ten variables, although they did not study the effect of these variables at the provincial level. Finally, Babula and Zhang (2019) adopted a cointegrated VAR model with a Cobb-Douglas production function to further study the influence of the housing sector on the demand for lumber in the U.S.

These studies provided either a model for studying U.S. demand for lumber in relation to Canadian export prices, or a model of Canada's export supply function to the U.S.; the models tend to consider domestic U.S. variables, Canada–U.S. exchange rates, and U.S. lumber prices. However, dynamic models that simultaneously consider demand and supply factors in both countries are lacking. A new approach to better understand the spillovers and the feedback between countries is the Global Vector Autoregressive (GVAR) approach.

In this chapter, we introduce and evaluate the innovative GVAR econometric method initially proposed and developed by Pesaran *et al.* (2004) and Dees *et al.* (2007). The GVAR approach had been used to examine trade in commodity (oil and wheat) markets as well as financial markets. Several papers examined the propagation mechanism of an adverse oil supply shock on the macroeconomic variables of oil importing and exporting countries (Dees *et al.*, 2007; Chudik and Fedora, 2012; Mohaddes and Pesaran, 2016). Gutierrez *et al.* (2015) examined the link between food and energy prices using a GVAR framework; their impulse responses indicated a negative impact of an increase in oil prices and real exchange rates on wheat prices.

The GVAR approach is appealing since it is designed to model fluctuations and interactions between countries/regions at the international level and, in addition, it enables us to model the dynamics of the prices associated with country-specific and/or foreign-specific variables. Unlike structural models, such as SPE trade models, GVAR does not attempt to enforce constraints associated with economic theory. Rather, the causal relationships are analyzed using impulse response functions to highlight how shocks on prices, exchange rates, housing starts, and global oil prices propagate at the domestic and global level.

The chapter is organized as follows. We begin in section 8.1 with a description of the GVAR approach, followed in section 8.2 by a description of the data we employ. The estimation results and discussion are found in section 8.3. We also investigate respective one standard error shocks to American housing starts, a reduction in real GDP, and an oil-price shock in section 8.4. Also, to mimic the potential effect of the COVID-19 pandemic, we examine an extreme shock to global GDP in section 8.4. Finally, some concluding comments are provided in section 8.5.

8.1 Methodology: Global Vector Autoregressive Models

A GVAR model is constructed in three stages. In the first stage, for each country or region a VAR model is developed including domestic variables as in a conventional VAR model. Moreover, each country is connected to the others using a set of foreign-specific variables. The foreign variables are assumed to be weakly exogenous and are constructed as weighted averages of the trade flows of their trading partners. These variables play an important role in measuring the influence of the rest of the world on each country/region, such as the influence of a shock in U.S. housing starts on Alberta's softwood lumber sector. The estimation stage of the reduced form of the VAR model is augmented by the exogenous variables and denoted by VARX. In the second step, the VARX models for individual countries are joined together and estimated as a GVAR model. Finally, the dynamic properties of the model are analyzed through the propagation of shocks in the economy; these are referred to as Generalized Impulse Response Functions.

Suppose there are N countries/regions in the global economy and consider that the country- or region-specific variables are related to the global variables measured as country-specific weighted averages of foreign variables, plus deterministic variables (such as time trends) and global weakly exogenous variables (such as oil prices). We model each country as a *VARX(p,q)*:

$$\Omega_i(L, p_i)\, y_{it} = a_{i0} + \Theta_i(L, p_i)\, y^*_{it} + \Lambda_i(L, p_i) d_i + \varepsilon_{it}, \tag{8.1}$$

where the index $i = 1, \ldots, N$ refers to region or country; $t = 1, \ldots, T$ refers to time; a_{i0} is a $(k_i \times 1)$ vector of deterministic intercepts; y_{it} is a $(k_i \times 1)$ vector of country-specific (domestic) variables and corresponding $(k_i \times k_i)$ matrices of lagged coefficients, respectively denoted by $\Theta_i(L, p_i)$ and $\Lambda_i(L, p_i)$; d_i represents the global exogenous variables; and ε_{it} is a $(k_i \times 1)$ matrix of idiosyncratic shocks assumed to be serially uncorrelated with mean zero and time-invariant covariance matrix Σ_{ii}. To estimate *VARX(p,q)*, the lag order of the polynomial matrices $\Omega_i(L, p_i)$, $\Theta_i(L, p_i)$ and $\Lambda_i(L, p_i)$ has to be determined. Pesaran *et al.* (2004) and other studies employ the Akaike information criterion (AIC) and the Schwarz Bayesian criterion (SBC) to find the lag orders.

After estimating each of the country-specific models, we can construct a GVAR model. Suppose we stack domestic and foreign specific variables in a $(k_i + k_i^*) \times 1$ matrix: $z_{it} = \begin{pmatrix} y_{it} \\ y^*_{it} \end{pmatrix}$. Then, the VARX model for each country is

$$B_{i0}\, z_{it} = a_{i0} + a_{i1} t + B_{i1}\, z_{i,t-1} + \varepsilon_{it}, \tag{8.2}$$

where $B_{i0} = (I_{k'}, -\Theta_{i0})$ and $B_{i1} = (\Omega_{i1}, \Theta_{i1})$, and I_k is a $(k \times k)$ identity matrix. In the next step, all country-specific variables are collected in a $(k^* \times 1)$ global vector defined as follows:

$$y_t = \begin{pmatrix} y_{0t} \\ y_{1t} \\ \vdots \\ y_{Nt} \end{pmatrix}.$$

The country-specific variables can now be written in terms of y_t using the weighting matrix:

$$z_{it} = W_i\, y_t, i = 0, 1, 2, \ldots, N, \tag{8.3}$$

where W_i is the link matrix constructed from the export weights that allow the country-specific models to be written in terms of the global vector, y_t. Hence, we have

$$B_{i0}\, W_i\, y_{it} = a_{i0} + a_{i1}t + B_{i1}\, W_i\, y_{i't-1} + \varepsilon_{it} \tag{8.4}$$

By stacking the country-specific model in equation (8.4), we get:

$$H_0\, y_{it} = a_{i0} + a_1 t + H_1\, y_{i't-1} + \varepsilon_{it}, \tag{8.5}$$

where the set of endogenous variables for country i includes:

$$H_0 = \begin{pmatrix} B_{00}W_0 \\ B_{10}W_1 \\ \vdots \\ B_{N0}W_N \end{pmatrix} H_1 = \begin{pmatrix} B_{01}W_0 \\ B_{11}W_1 \\ \vdots \\ B_{N1}W_N \end{pmatrix} a_0 = \begin{pmatrix} a_{00} \\ a_{10} \\ \vdots \\ a_{N0} \end{pmatrix} a_1 = \begin{pmatrix} a_{01} \\ a_{11} \\ \vdots \\ a_{N1} \end{pmatrix} \varepsilon_t = \begin{pmatrix} \varepsilon_{0t} \\ \varepsilon_{1t} \\ \vdots \\ \varepsilon_{Nt} \end{pmatrix}.$$

Assuming H_0 is non-singular, we can multiply both sides of equation (8.5) by H_0^{-1} and write the GVAR model as follows:

$$y_t = H_0^{-1}\, a_0 + H_0^{-1}\, a_1 t + H_0^{-1}\, H_1\, y_{t-1} + H_0^{-1}\, \varepsilon_{it}. \tag{8.6}$$

Equation (8.6) can be solved by means of a recursive method and then used to analyze the impulse responses and make forecasts. A comprehensive version of the GVAR model can be found in Pesaran *et al.* (2004, 2009) and di Mauro and Pesaran (2013).

8.2 Data

The version of the GVAR model developed in this chapter involves six regions/countries. We divide Canada into five regions – British Columbia (BC); Alberta (AB); Eastern Canada (EC), which includes Ontario and Quebec; Atlantic Canada (AC), which includes Newfoundland and Labrador; New Brunswick, Prince Edward Island and Nova Scotia; and the rest of Canada (ROC), which includes Saskatchewan and Manitoba. Since our focus is on regional impacts on Canada's various lumber producing regions, the U.S. is considered as a single region. Models are estimated using quarterly data from 1996Q1 to 2017Q4; these data cover the latest trade agreements discussed in Chapter 2 (SLA1996 and SLA2006). The variables used in the model are the U.S.–Canada exchange rate (CAN\$ per US\$), the long-term nominal interest rate in North America, real gross domestic products (measured in billions of dollars in local currency), total housing starts in North America ('000s of units, not seasonally adjusted), export prices of softwood lumber from Canadian regions to the U.S. (CAN\$/m^3), an index of the U.S. price of lumber (US\$/m^3), and the end-of-period softwood lumber inventory held by Canadian lumber manufacturers ('000s m^3) (see Jennings *et al.*, 1991). Finally, oil prices are considered to be a global variable.

The U.S. softwood lumber price index is converted to a 2012 basis and comes from the Federal Reserve Bank of St Louis (FRED) (2020). Data on Canadian end-of-year inventory, real GDP, and export prices are from Statistics Canada (2019a,b,c). Ten-year government bond rates are used for long-term interest rates and, along with exchange rates, are also from the FRED. The U.S.

real GDP is from the U.S. Bureau of Economic Analysis (2019) and housing starts are from the U.S. Census Bureau (2019a,b,c).

The data are transformed into natural logs and converted to real terms where appropriate:

$$y_{it} = \ln(GDP_{it} / CPI_{it}) \qquad p^e_{it} = \ln\left(P^e{}_{it}\right)$$
$$hs_{it} = \ln(HS_{it}) \qquad ls_{it} = \ln\left(LS_{it}\right)$$
$$er_{it} = \ln(ER_{it} / CPI_{it}) \qquad lr_{it} = 0.25 \times \ln\left(1 + R_{it} / 100\right)$$
$$p^o{}_{it} = \ln(P^{oil}_t)$$

Here y_{it} is the natural logarithm of the real GDP of region *i* at time *t*; GDP_{it} is the nominal gross domestic product of region *i* at time *t*; CPI_{it} is the consumer price index in region *i* at time *t*; hs_{it} is natural logarithm of housing starts in North America; $p^e{}_{it}$ is the natural logarithm of lumber prices in region *i*; ls_{it} is the natural logarithm of the end-of-year inventory of lumber; R_{it} is the nominal rate of interest (per annum in percent); er_{it} is the real exchange rate; and $p^o{}_t$ is natural logarithm of oil prices – the updated Brent crude oil price index measured in U.S. dollars (Mohaddes and Raissi, 2018).

The foreign-specific variables are defined as the weighted average of domestic variables and denoted with a star (*); for example, $y^*{}_{it}$ is a ($k \times i \times 1$) matrix that is constructed using weighted averages of the region-specific weights. The average of real GDP is given as $y^*{}_{it} = \sum_{j=0}^{N} w_{ij} y_{jt}$; the average of lumber prices, $p^{e*}{}_{it} = \sum_{j=0}^{N} w_{ij} p^e{}_{jt}$; the average of housing starts, $hs^*{}_{it} = \sum_{j=0}^{N} w_{ij} hs_{jt}$; the average of end-of-period inventory, $ls^*{}_{it} = \sum_{j=0}^{N} w_{ij} ls_{jt}$; and the average of long-term interest rates, $lr^*{}_{it} = \sum_{j=0}^{N} w_{ij} lr_{jt}$.

Quantitatively, the weights are measured as the average of total merchandise trade. We employed fixed weights over time using average 2001–2012 regional trade data to reduce the effect of the individual yearly movement in trade. More generally, the trade weights are computed as follows:

$$w_{ij} = \frac{Tr_{ij,2010} + Tr_{ij,2011} + Tr_{ij,2012}}{Tr_{i,2010} + Tr_{i,2011} + Tr_{i,2012}},$$

where Tr_{ij} represents the total trade between regions and w_{ij} is the average weight. The weight matrix produces interesting information on cross-region linkages and highlights the dominant role of the main trading partner for each region in the model (Table 8.1). Weights could also vary over time to capture the changes in trade patterns, although frequent changes in the weight matrix may lead to unfavorable randomness in the data analysis (see Pesaran *et al.*, 2004). We assume the foreign-specific variables are weakly exogenous since most of our regional economies (except the U.S.) are relatively small (Smith and Galesi, 2014).

The region-specific variables are as follows:

Canadian regions: $(y_{it}, p^e{}_{it}, lr_{it}, ls_{it}, er_{it}, hs_{it}), i = 1, \ldots, 5;$

United States: $(y_{it}, p^e{}_{it}, lr_{it}, hs_{it}).$

Table 8.1. Trade weight[a] based on total merchandise exports. (Adapted from U.S. International Trade Data (2019), Statistics Canada (2019a), and authors' calculations.)

Variables	BC	AB	ROC	EC	AC	U.S.
Alberta	0.061	0.000	0.061	0.225	0.018	0.130
British Columbia	0.000	0.047	0.023	0.158	0.008	0.103
Rest of Canada	0.015	0.052	0.000	0.119	0.011	0.059
Eastern Canada	0.068	0.072	0.105	0.000	0.125	0.636
Atlantic Canada	0.007	0.008	0.007	0.143	0.000	0.072
U.S.	0.849	0.822	0.805	0.355	0.839	0.000

[a] Trade weights are computed as shares of exports and imports in rows by region from 2009 to 2012. The weights are displayed in columns by export country and the columns sum to one.

Note that the real exchange rate is not included in the U.S. VARX model. The data for the end-of-year lumber stocks are not publicly available for the U.S., while Canadian regions face the same exchange rate. Accordingly, the set of foreign variables included in each country's regions are as follows:

Canadian regions: $(y^*_{it,} p^{e*}_{it,} lr^*_{it,} hs^*_{it,}), l = 1, \ldots, 5;$

U.S. regions: $(y^*_{it,} p^*_{it,} er^*_{it,} hs^*_{it}).$

Finally, the global variable p^o_{it} pertains to all regions.

8.3 Estimation of the GVAR Model

Following Pesaran *et al.* (2004), we assume the country-specific variables are non-stationary ($I(1)$). The Augmented Dickey-Fuller (ADF) test statistics are used where the null hypothesis is the non-stationarity of the variables. To run the test, we assume the equations first involve intercepts and linear trends for the variables in levels and first-differenced form; and, second, involve the intercept only. We restricted the maximum number of lags for the domestic and foreign variables to four, as data are quarterly. The asymptotic 5% critical values are used. Table 8.2 represents the results of the ADF test. We reject the null-hypothesis that the variables are non-stationary at the level with trend or no trend.

We use AIC and SBC to estimate the lag length of the VARX(p,q) for the individual models. We assume $p_i \geq q_i$ to control the total dimensions of the system with maximum length of three for domestic variables and two for foreign variables (Pesaran *et al.*, 2009). The VARX model are defined using the highest values of AIC and SBC tests. The rank of the cointegrating space for the individual models are computed both for the trace and maximum eigenvalues statistics (Pesaran *et al.*, 2004). The results are reported in Table 8.3.

The cointegration tests are conducted using both trace and maximum eigenvalue statistics to determine the cointegrating space for individual models. The values of the test statistics and the critical values are reported in Table 8.4.

Table 8.2. Augmented Dickey-Fuller unit root statistics for home and foreign variables[a]. (Author's own table.)

Regions	AB	BC	ROC	EC	AC	U.S.
y_{it}	–0.919	0.338	–0.855	–1.459	0.615	–1.138
p_{it}	–0.251	–0.024	–0.478	–0.955	–0.829	–2.176
lr_{it}	–0.694	–0.694	–0.694	–0.694	–0.694	–1.627
er_{it}	–1.667	–1.667	–1.667	–1.667	–1.667	—
hs_{it}	–2.883	–3.166	–1.714	–1.554	–2.073	–0.938
lst_i	–2.984	–2.698	–2.088	–2.022	–2.237	—
y^*_{it}	–0.953	–1.239	–1.054	–0.183	–1.313	–1.086
p^*_{it}	–1.766	–1.665	–1.201	–0.188	–1.388	–0.448
lr^*_{it}	–1.408	–1.371	–1.348	–1.167	–1.393	–0.694
er^*_{it}	–1.667	–1.667	–1.667	–1.667	–1.667	–1.667
hs^*_{it}	–0.873	–0.880	–0.871	–1.201	–0.862	–1.607
lst^*_{it}	–2.194	–2.131	–2.464	–2.262	–2.165	–2.232
$poil^*$	–2.050	–2.050	–2.050	–2.050	–2.050	–2.050

[a] The ADF test statistics is computed based on a univariate AR(p) model in the levels including an intercept. The 95% critical value of the ADF statistic for a regression with a constant are –2.89 and –3.15.

Table 8.3. VARX order and number of cointegration relationship in the country-specific models[a]. (Author's own table.)

	VARX*(p_i, q_i)		
Region	p_i	q_i	Number of cointegrated relationships
AB	3	2	3
BC	3	1	4
ROC	3	2	2
EC	3	2	2
AC	2	1	2
US	3	2	1

[a] Rank orders are the result of GVAR program using Johansen's trace test at 5% significant level.

The critical values at the 5% level of significance are taken from Mackinnon *et al.* (1999). Overall, both test statistics result in the same rank orders and, if there is a difference, we relied on the trace test statistics (Smith and Galesi, 2014). The trace test statistics provide better results on cointegration when the sample is small compared with the maximum eigenvalue statistics. All regions have acceptable results.

After finding the lag orders for domestic and foreign variables, we can proceed to estimate updated versions of the VARX models, referred to as Vector Error-Correcting models (VECM) (for further details, see Pesaran *et al.*, 2004). The updated models are estimated using the endogenous variables (domestic variables with lag order p_i) and the weakly exogenous variables (foreign variables with lag order of q_i-). The GVAR model is built on the important assumption that foreign variables are weakly exogenous, so no region could have a long-term influence

Table 8.4. Cointegration rank statistics results[a]. (Author's own table.)

Region	Trace test				Maximum eigenvalue test			
Alberta	H0	H1	statistic	95% crit.v.	H0	H1	statistic	95% crit.v.
	r=0	r=1	278.88	184.53	r=0	r>1	67.84	63.52
	r<1	r=2	184.11	145.3	r<1	r≥2	62.60	57.13
	r≤2	r=3	118.94	110.03	r≤0	r≥3	32.99	50.64
	r≤3	r=4	77.61	78.52	r≤0	r≥4	29.66	43.94
	r≤4	r=5	40.92	50.72	r≤0	r≥5	22.03	36.84
	r≤5	r=6	13.68	26.24	r≤0	r≥6	16.71	28.81
British Columbia	r=0	r=1	250.55	184.53	r=0	r>1	102.75	63.52
	r<1	r=2	182.04	145.3	r<1	r≥2	52.46	57.13
	r≤2	r=3	126.43	110.03	r≤0	r≥3	47.42	50.64
	r≤3	r=4	81.61	78.52	r≤0	r≥4	26.99	43.94
	r≤4	r=5	47.11	50.72	r≤0	r≥5	25.52	36.84
	r≤5	r=6	19.92	26.24	r≤0	r≥6	20.57	28.81
Rest of Canada	r=0	r=1	239.08	184.53	r=0	r>1	92.51	63.52
	r<1	r=2	157.03	145.3	r<1	r≥2	49.99	57.13
	r≤2	r=3	101.94	110.03	r≤0	r≥3	43.51	50.64
	r≤3	r=4	56.53	78.52	r≤0	r≥4	27.44	43.94
	r≤4	r=5	29.93	50.72	r≤0	r≥5	22.57	36.84
	r≤5	r=6	13.03	26.24	r≤0	r≥6	17.82	28.81
Eastern Canada	r=0	r=1	264.63	184.53	r=0	r>1	89.32	63.52
	r<1	r=2	157.54	145.3	r<1	r≥2	55.78	57.13
	r≤2	r=3	108.90	110.03	r≤0	r≥3	47.67	50.64
	r≤3	r=4	67.67	78.52	r≤0	r≥4	32.83	43.94
	r≤4	r=5	36.00	50.72	r≤0	r≥5	26.75	36.84
	r≤5	r=6	12.70	26.24	r≤0	r≥6	14.67	28.81
Atlantic Canada	r=0	r=1	265.20	184.53	r=0	r>1	117.99	63.52
	r<1	r=2	162.24	145.3	r<1	r≥2	57.62	57.13
	r≤2	r=3	101.17	110.03	r≤0	r≥3	44.98	50.64
	r≤3	r=4	60.45	78.52	r≤0	r≥4	30.99	43.94
	r≤4	r=5	31.89	50.72	r≤0	r≥5	24.92	36.84
	r≤5	r=6	9.20	26.24	r≤0	r≥6	16.74	28.81
United States	r=0	r=1	168.77	171.33	r=0	r>1	87.73	57.38
	r<1	r=2	88.22	134.16	r<1	r≥2	51.21	51.1
	r≤2	r=3	52.19	100.96	r≤0	r≥3	30.97	44.71
	r≤3	r=4	27.32	71.56	r≤0	r≥4	29.42	38.15
	r≤4	r=5	10.80	45.9	r≤0	r≥5	16.26	31.28

[a] The null hypothesis tests the cointegration vectors r against the alternative hypothesis of (maximum) r+1 cointegration vectors for a maximum eigenvalue (trace) test. The statistical values are at the 5% significance level and are taken from Mackinnon *et al.* (1999).

on the market but only a short-term one. To test for the weak exogeneity, the Johansen (1992) and Harbo *et al.* (1998) tests are used. Table 8.5 indicates that the null hypothesis of weak exogeneity for foreign-specific variables cannot be rejected since the F-statistics calculated from the data are smaller than the critical value for the majority of foreign and global variables. Therefore, the GVAR model can be used to compute the impulse response functions of the country/ region and the foreign specific variables.

We can also analyze the contemporaneous effect of foreign variables on their corresponding domestic variables. Table 8.6 provides estimated coefficients of the contemporaneous effects between domestic and foreign variables and their associated t-statistics. These coefficients describe the short-run relationships and the strength of the linkages between foreign and domestic variables. We find that all the coefficients are positive and mostly significant for all regressors. Housing starts have a positive and significant impact in all regions, with Eastern Canada and the United States having higher impact elasticities. Positive and significant effects are provided by real GDP in all regions, while elasticities for Eastern Canada and the Rest of Canada exceed 1.0. Looking at the prices of lumber, we find all estimations are significant except for the U.S. market, suggesting a poor relationship between domestic and foreign lumber prices. Lastly, the t-statistics for the long-term interest rate are insignificant in the majority of regions.

Table 8.5. F-statistics for testing the weak exogeneity of country-specific foreign and global variables[a]. (Author's own table.)

Region	y^*_{it}	p^*_{it}	hs^*_{it}	lr^*_{it}	er^*_{it}	P^o_{it}
Alberta	0.718*	0.824*	0.938*	6.272	—	1.361*
British Columbia	0.942*	0.561*	2.092*	1.196*	—	1.719*
Rest of Canada	1.930*	0.418*	0.001*	0.214*	—	0.521*
Eastern Canada	6.421	0.709*	9.551	0.038*	—	1.117*
Atlantic Canada	2.761*	1.010*	9.203	0.755*	—	0.850*
United States	2.230*	0.452*	0.372*	—	1.891*	—

[a] The null hypothesis test is that the foreign variables are weakly exogenous. The * indicates that the null hypothesis cannot be rejected at the 5% level of significance

Table 8.6. Contemporaneous effects of foreign variables on home-specific counterparts[a]. (Author's own table.)

Region	y^*_{it}	p^{e*}_{it}	hs^*_{it}	lr^*_{it}
Alberta	0.597	3.468	0.241	0.308
	(4.021)	(6.724)	(6.083)	(2.124)
British Columbia	0.880	2.676	0.205	0.118
	(4.908)	(8.615)	(5.345)	(1.013)
Rest of Canada	1.089	3.084	0.283	0.147
	(5.542)	(8.816)	(7.248)	(1.001)
Eastern Canada	1.472	1.116	0.689	1.209
	(5.616)	(11.217)	(5.401)	(14.697)
Atlantic Canada	0.428	3.138	0.283	0.173
	(2.257)	(8.274)	(6.667)	(1.082)
United States	0.192	0.042	0.607	-
	(2.712)	(1.922)	(5.070)	-

[a] The robust t-statistics in the parentheses are computed using White's heteroscedastic corrected standard errors.

8.4 Generalized Impulse Response Analysis

In this section, we analyze the generalized impulse response functions (GIRF) suggested by Koop *et al.* (1996), and by Pesaran and Shin (1996), to explore the short-run dynamics. We are interested in the effect that changes in the U.S. economy will have on Canadian softwood lumber; therefore, we focus on the Canadian regions' GIRF in response to several shocks. To quantify the dynamics of the GVAR model and the time path of the effects of the shocks to domestic and foreign variables, we consider the implications of the following shocks over 40 quarters:

- A one standard error positive shock to U.S. housing starts;
- A one standard error negative shock to U.S. real GDP;
- A COVID-19 (three standard error negative) shock to U.S. real GDP; and
- A one standard error positive shock to oil prices.

First, we analyze how a positive shock to the U.S. housing market influences export prices of lumber across Canada. A one standard deviation shock is equivalent to a 3.7% increase in U.S. housing starts. Figure 8.1 indicates the General Impulse Responses with a solid line and 90% bootstrapped confidence intervals by dashed lines. Export prices in Canada are obtained by dividing the value of lumber exports to the U.S. by lumber quantity. Thus, the export price response might include the effect of housing starts on the quantity of lumber exported and its value, or both.

For the purpose of this chapter, we study price responses to various shocks that are the main drivers in international forest products trade disputes. In the case of the U.S.–Canada dispute, a positive shock in the U.S. housing market (Fig. 8.1) increases the demand for softwood lumber from Canada. The response to this shock is an increase in lumber export prices from AB, BC and ROC by 4.6% in the second quarter, and increases of 6.7% and 1.7% in the prices of lumber exports from AC and EC, respectively. The maximum impact is on Atlantic Canada (AC) where New Brunswick, which is exempt from export duties, is the main exporter on softwood lumber. These results correspond with actual observations. After the immediate impact, however, longer-term impacts oscillate, dissipating over time but remaining well above zero.

U.S. real GDP is one of the main macroeconomic factors that has a direct impact on the U.S. demand for Canadian lumber and, accordingly, lumber prices. As indicated in Fig. 8.2, a negative shock to U.S. real GDP (–0.69%) results in less than a 1% increase in lumber prices in the first quarter in AB, BC, EC, AC, and the ROC, followed by a sharp decline starting from the second quarter. Prices in AB start to increase after four quarters and then continue to fluctuate, while remaining below zero. Lumber prices start to increase slightly after the 4th quarter in BC and AC, but remain relatively stable over time and slightly above zero. The price fluctuations in the ROC and EC follow the same pattern, converging to almost no change in future quarters.

Next, in Fig. 8.3, we consider a sharp decline in U.S. real GDP of three standard deviations, which mimics a massive decline in U.S. demand due to the COVID-19 pandemic. On impact in the first quarter, the COVID-19 shock results in a 2.9% increase in Alberta prices of lumber, 4.7 % in BC, 2.3% in EC, 4.6% in AC, and 2% in the ROC. The higher lumber prices continue into the

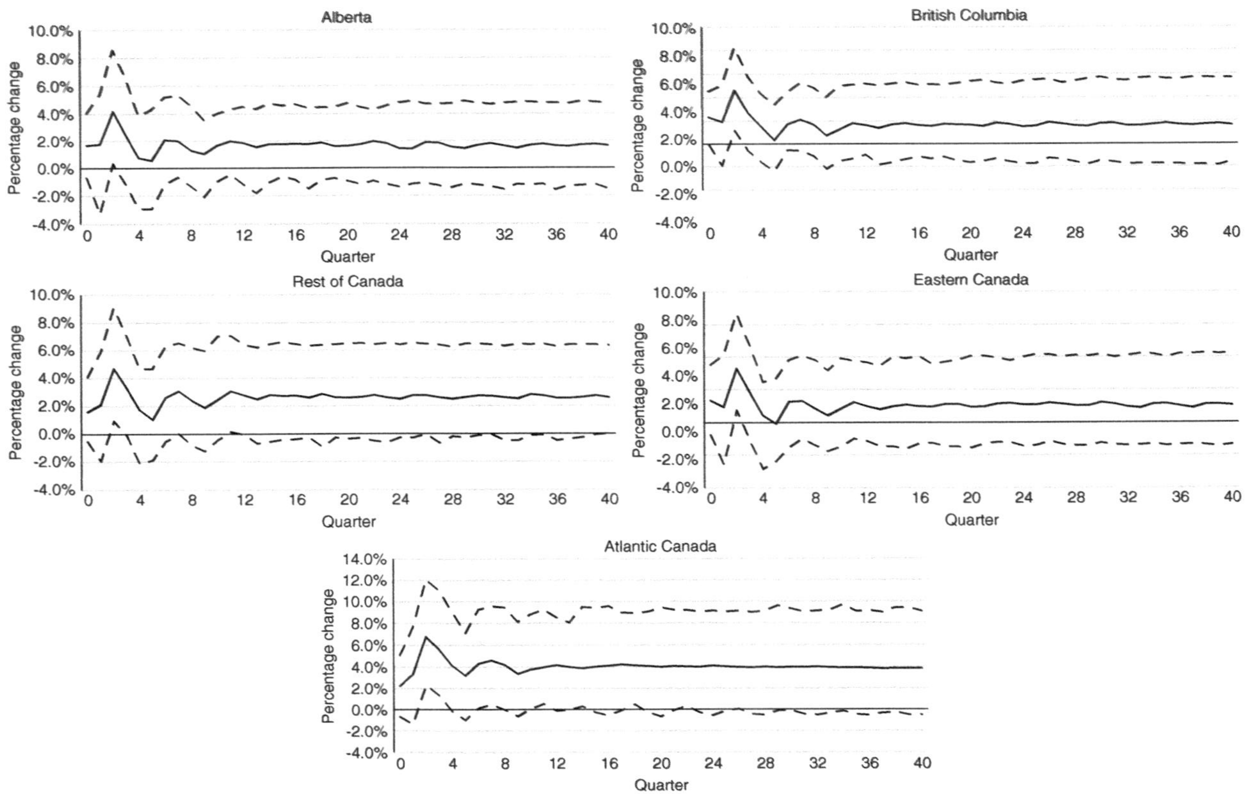

Fig. 8.1. Generalized impulse responses of a positive one standard error shock to the U.S. housing starts on lumber export prices in Canadian regions. (Author's own figure.)

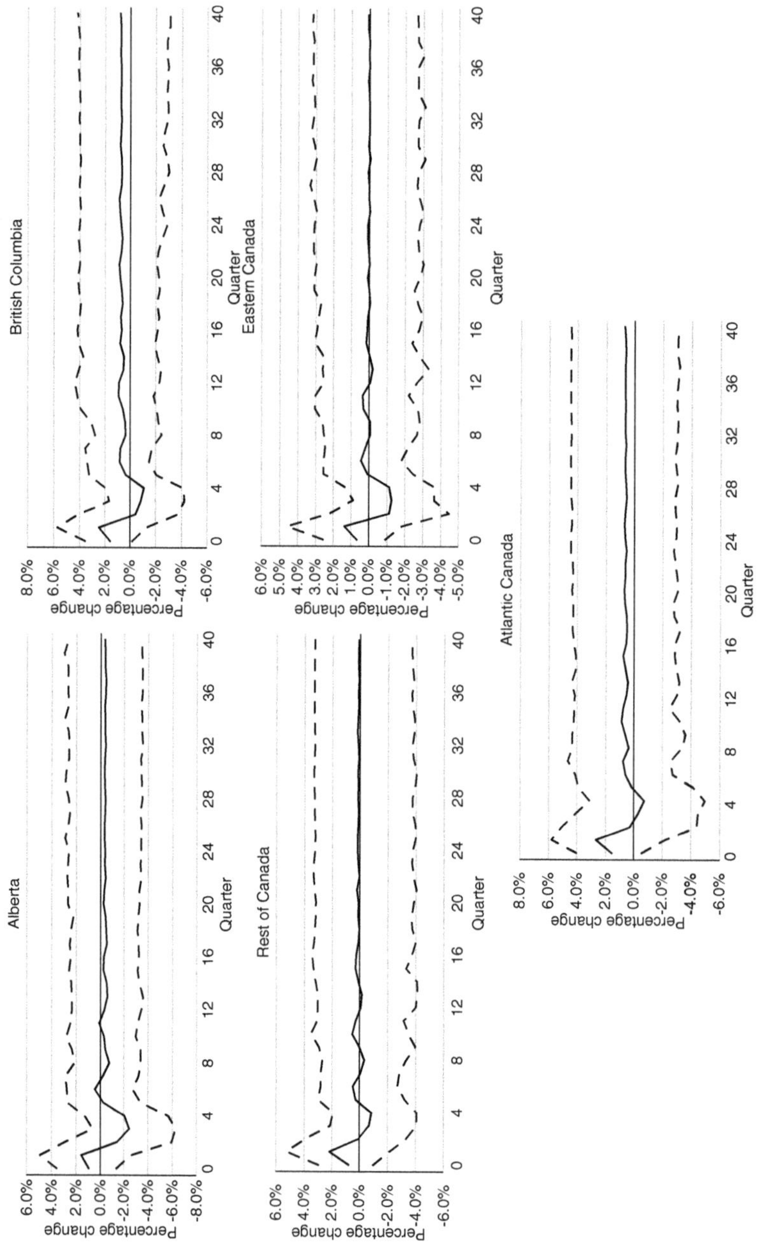

Fig. 8.2. Generalized impulse responses of a negative one standard error shock to the U.S. real GDP on lumber export prices in Canadian regions. (Author's own figure.)

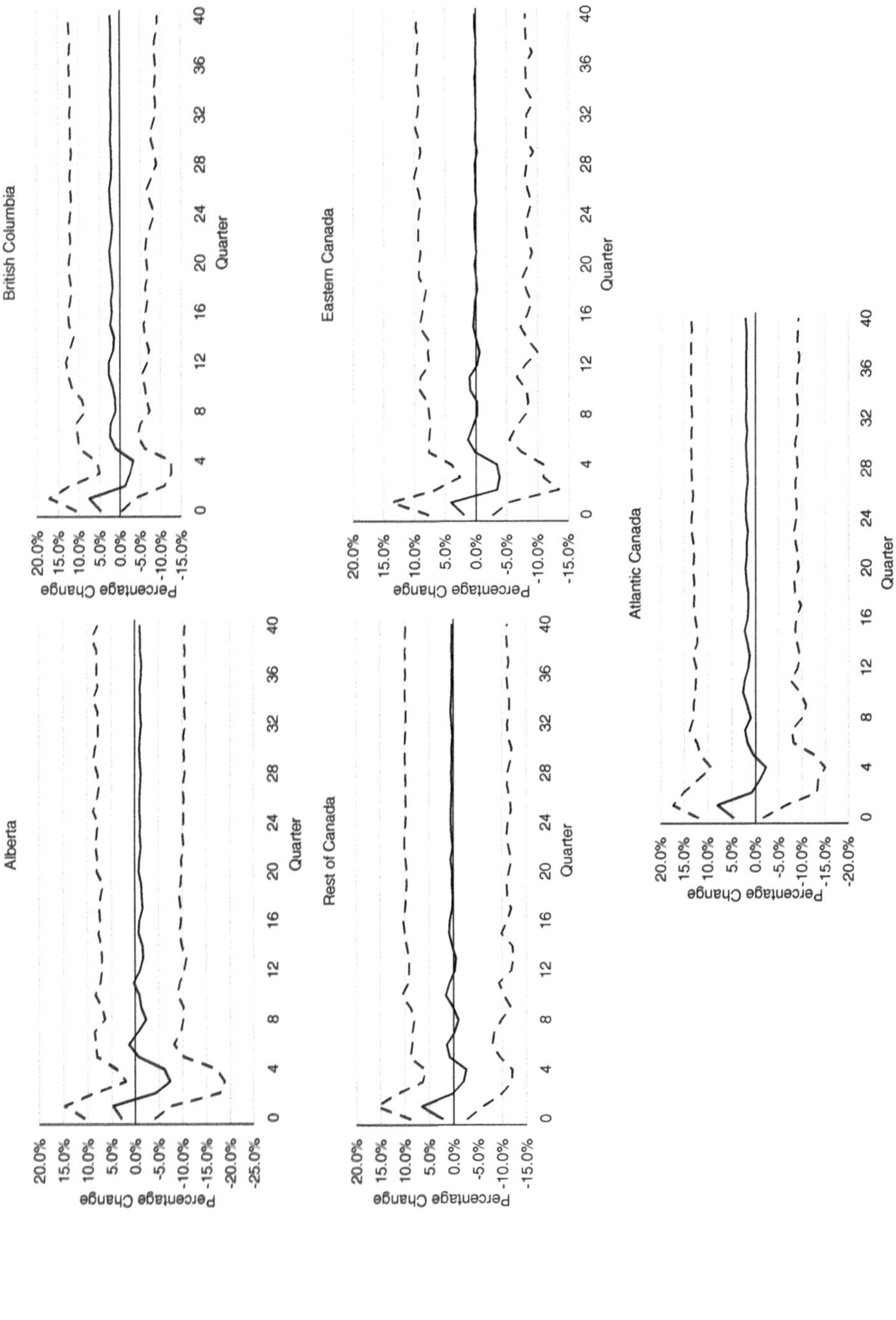

Fig. 8.3. Generalized impulse responses of a negative three standard error shock to the U.S. real GDP on lumber export prices in Canadian regions. (Author's own figure.)

second quarter, with the greatest increase in BC of 7.5% and the least in the ROC at 2%; this suggests that the price increase is due to a shortage in supply that exceeds the impact of a fall in demand, with the market responding differently across provinces. After two quarters, however, the market adjusts to the huge decline in U.S. real GDP. The large drop in the demand for products in the U.S. encompasses all goods, including softwood lumber from Canada; prices in the third quarter then fall by 9.0% in Alberta, 8.7% in BC, 6.3% in EC, 7.1% in AC, and 7.5% in the ROC. Prices in Alberta start to increase slightly after four quarters and continue to fluctuate thereafter, while remaining below the pre-COVID level. Lumber prices start to increase slightly after the 4th quarter in BC and the ROC as well, but remain relatively stable over time and only slightly above pre-COVID prices. The price fluctuations in EC and the ROC follow the same pattern converging to almost no change in future quarters.

Finally, we study the effect of a global oil price shock on lumber export prices in various Canadian regions. In Fig. 8.4, we assume a one standard deviation positive shock to the oil price – an increase of 11% in the oil price index in one quarter. Not surprisingly, the impact of an oil price shock is the highest in Alberta, increasing the price of lumber by 5.6% compared with other provinces. For the remaining regions the impact is equal to 2.8%–4.2%. Interestingly, the propagation mechanism varies significantly for Alberta compared with other regions. While the export price in Alberta stays above what it would be in the absence of an oil price shock, the export prices in other regions fall below pre-shock prices beyond the fourth quarter. One possible explanation is that positive oil prices increase aggregate demand in Alberta as well as more resource revenue for the government, with the increase in provincial GDP translating into higher lumber prices. In the other provinces, a higher price of oil translates into lower lumber prices. Overall, as shown in Fig. 8.4, a sudden increase in the price of oil has a significant effect on lumber prices.

8.5 Concluding Remarks

The purpose in this chapter was to introduce a new approach to studying the softwood lumber trade market. We employed the Global Vector Autoregressive (GVAR) method to assess and investigate variables that are key to the softwood lumber markets. In addition, we assessed the impact of global shocks, including an oil price shock and shock due to the COVID-19 pandemic. Our findings show strong co-movement in the export prices in various Canadian regions in response to changes in U.S. housing starts, with the greatest shocks occurring in Atlantic Canada and the least in Eastern Canada. It also turns out that softwood lumber export prices are quite sensitive to shocks in U.S. real GDP. Looking at the generalized impulse response functions, export prices rapidly drop below zero with a less than 1% reduction in the U.S. GDP. In the case of the COVID-19 shock, Canadian export prices fall by between 6.3% (Eastern Canada) and 9.0% (Alberta) by the third quarter after the shock occurs. As indicated in Appendix 8.A, a shock to long-term interest rates in the United

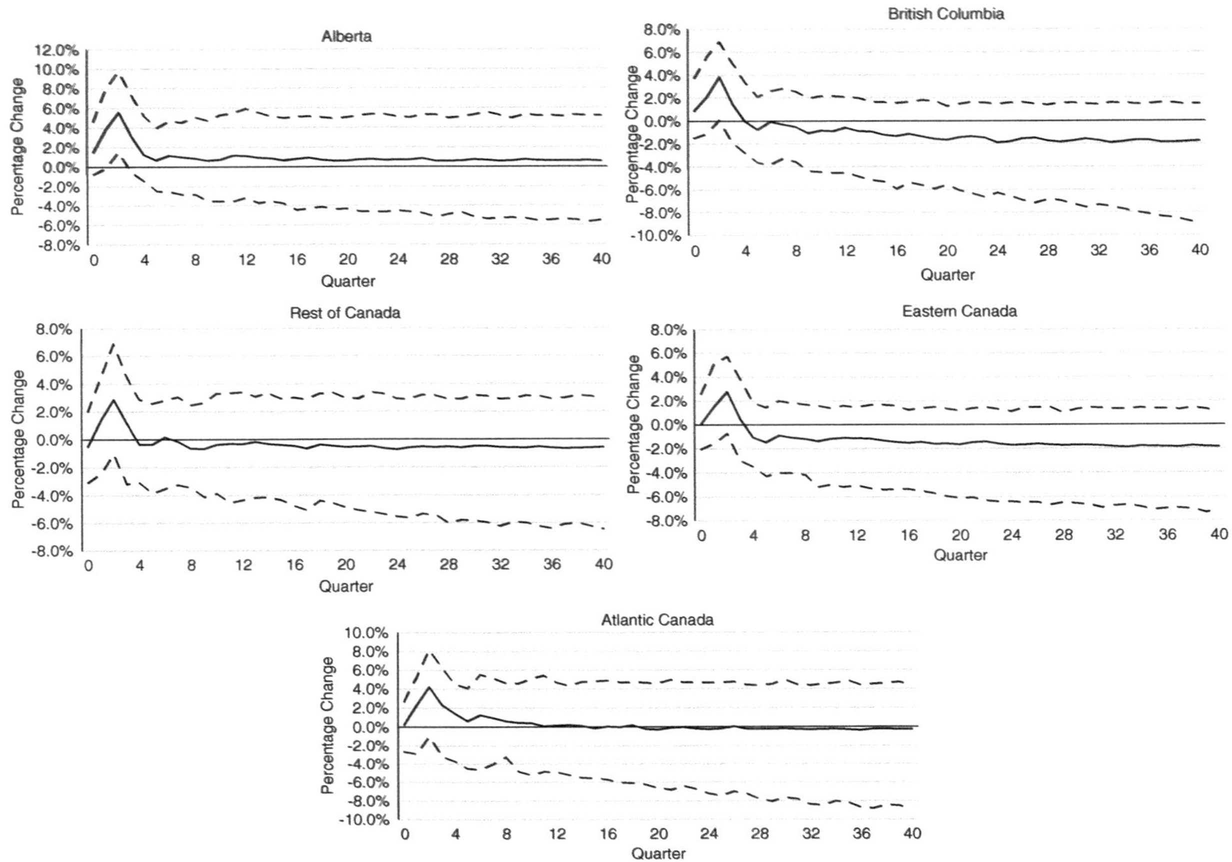

Fig. 8.4. Generalized impulse responses of a positive one standard error shock to oil prices on lumber export prices in Canadian regions. (Author's own figure.)

States, which might increase several periods after recovery from the COVID-19 pandemic, also has a notable impact on export prices in Canada's regions; such a shock may take upwards of eight quarters before prices revert to their pre-shock level. Meanwhile, the inflationary effect of a global shock in oil prices varies across Canada. While the prices respond positively in all regions, the propagation mechanism in Alberta significantly differs compared with other regions in Canada.

Future research is clearly needed to clarify the impacts of U.S. trade action on various regions in Canada, both in the near and longer terms. But research also needs to extend the GVAR approach to include other major exporters and importers of softwood lumber. It might be limited, for example, to country-level trade as opposed to trade between regions in Canada and the U.S., but it could provide the cross-linkage between countries to analyze the impact of shocks. In addition, the advantage of extending the analysis in this way relates to the importance of softwood lumber exporters such as Germany and the Scandinavian countries, which are able to expand their exports to the U.S. in response to U.S. duties and/or quotas on Canadian lumber. Finally, the model can be expanded to include export and import quantities separately in order to examine effects on lumber trade patterns.

Appendix 8.A: Descriptive Statistics and an Interest Rate Shock

Table 8.A.1. Domestic variables. (Author's own table.)

	REAL GDP		EXPORT PRICE		HOUSING STARTS		NOMINAL L RATE		REAL EXCHANGE RATE		LUMBER STOCK	
Domestic Variables	Mean	Std. dev.	Mean	Std. dev.	Mean	Std. dev.	Mean	Std. dev.	Mean	Std. dev.	Mean	Std. dev.
AB	13.46	0.80	5.09	0.27	8.93	0.33	0.04	0.02	–4.72	0.12	773	183
BC	12.81	0.64	5.24	0.23	8.80	0.36	0.04	0.02	–4.72	0.12	3344	603
ROC	14.38	0.51	5.13	0.21	10.17	0.30	0.04	0.02	–4.72	0.12	2890	562
EC	11.99	0.49	5.09	0.20	7.77	0.51	0.04	0.02	–4.72	0.12	341	90
AC	12.28	0.55	4.95	0.33	7.66	0.51	0.04	0.02	–4.72	0.12	113	53
U.S.	16.51	0.13	4.50	0.10	13.03	0.85	0.04	0.01				

Table 8.A.2. Foreign variables. (Author's own table.)

	REAL GDP		EXPORT PRICE		HOUSING STARTS		NOMINAL L RATE		REAL EXCHANGE RATE	
Foreign Variables	Mean	Std. dev.	Mean	Std. dev.	Mean	Std. dev.	Mean	Std. dev.	Mean	Std. dev.
AB	15.92	0.19	4.59	0.09	12.34	0.73	0.04	0.02		
BC	15.90	0.21	4.60	0.09	12.28	0.72	0.04	0.02		
ROC	15.71	0.21	4.62	0.09	12.10	0.70	0.04	0.02		
EC	14.34	0.45	4.90	0.17	10.32	0.36	0.04	0.02		
AC	15.83	0.19	4.60	0.08	12.24	0.73	0.04	0.02		
U.S.	12.41	0.54	5.10	0.21	8.16	0.40			–4.72	0.12

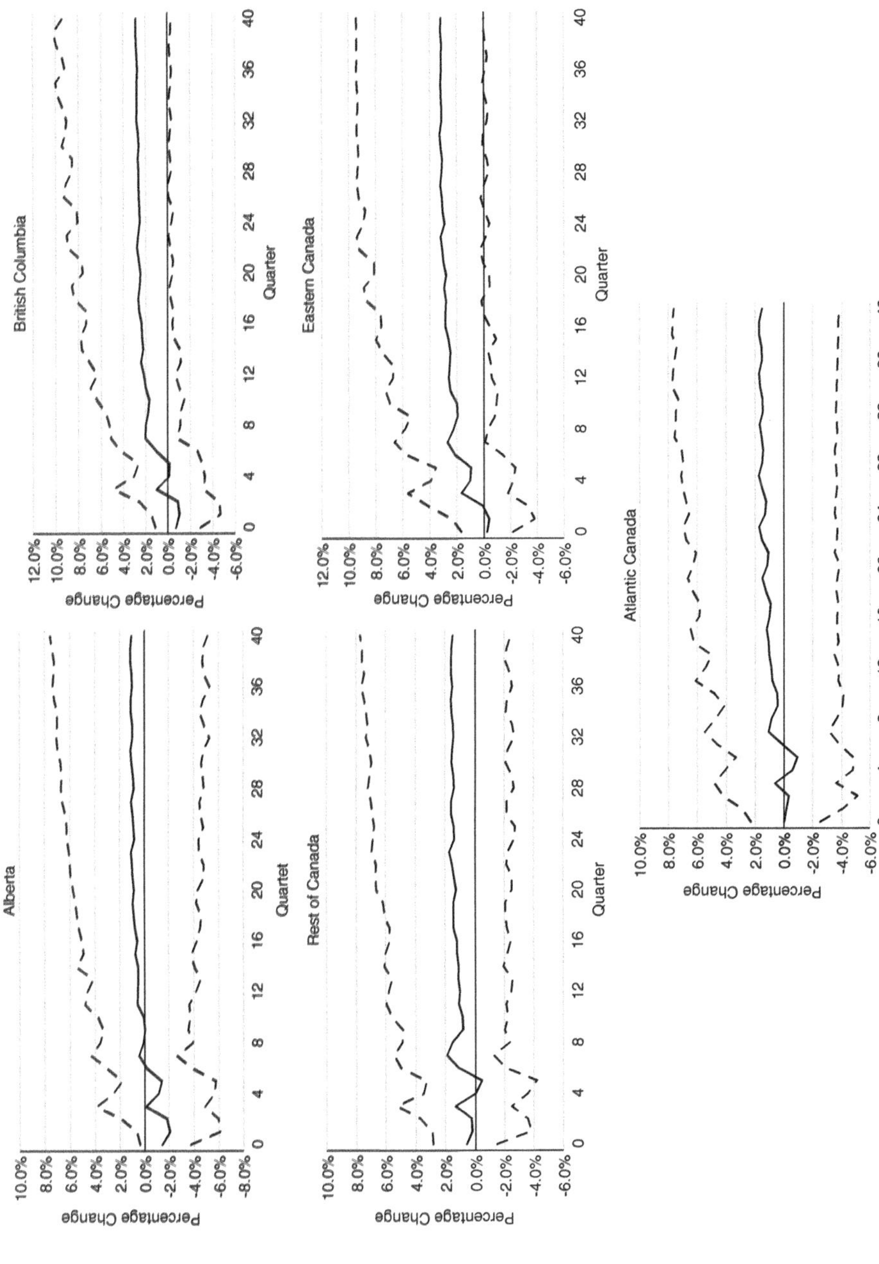

Fig. 8.A.1. Generalized impulse responses of a negative one standard error shock to the U.S. long-term interest rate on lumber export prices in Canadian regions. (Author's own figure.)

References

Babula, R.A. and Zhang, D. (2019) Assessing the role of futures position substitutability in a monthly model of factor demand for softwood lumber. *Empirical Economics* 56(4), 1097–1116.

Chudik, A. and Fedora, M.(2012) How the global perspective can help us to identify structural shocks. *Staff Paper No. 19*, Federal Reserve Bank of Dallas. https://www.dallasfed.org/~/media/documents/research/staff/staff1204.pdf (accessed July 20, 2020).

Dees, S., di Maouro, F., Pesaran, M.H. and Smith, L.V. (2007) Exploring the international linkages of the euro area: a global VAR analysis. *Journal of Applied Econometrics* 22(1), 1–38.

di Mauro, F. and Pesaran, M.H. (2013) *The GVAR Handbook: Structure and Applications of a Macro Model of the Global Economy for Policy Analysis*. Oxford University Press, Oxford, UK.

Federal Reserve Bank of St Louis, (2020) Producer Price Index by Commodity for Lumber and Wood Products: Logs, Bolts, Timber, Pulpwood and Wood Chips (WPU085), May. Available at: https://fred.stlouisfed.org/series/WPU085 (accessed June 11, 2020).

Gutierrez, L., Francesco, P. and Roggero, P.P. (2015) A global vector autoregression model for the analysis of wheat export prices. *American Journal of Agricultural Economics* 97(5), 1494–1511.

Harbo, I., Johansen, S., Nielsen, B. and Rahbek, A. (1998) Asymptotic inference on cointegrating rank in partial systems. *Journal of Business and Economic Statistics* 16, 388–399.

Jennings, S., Adamowicz, W. and Constantino, L. (1991) The Canadian lumber industry and the macroeconomy: a vector autoregression analysis. *Canadian Journal of Forest Research* 21, 288–299.

Johansen, S. (1992) Cointegration in partial systems and the efficiency of single-equation analysis. *Journal of Econometrics* 52, 231–254.

Koop, G., Pesaran, H.M. and Potte, S.M. (1996) Impulse response analysis in nonlinear multivariate models. *Journal of Econometrics* 74(1), 119–147.

MacKinnon, J., Haung, A. and Michelis, L. (1999) Numerical distribution functions of likelihood ratio tests for cointegration. *Journal of Applied Econometrics* 14(5), 563–577.

Mohaddes, K. and Pesaran, M.H. (2016) Country-specific oil supply shocks and the global economy: a counterfactual analysis. *Energy Economics* 59, 382–399.

Mohaddes, K. and Raissi, M. (2018) Compilation, revision and updating of the Global VAR (GVAR) Database, 1979Q2-2016Q4 (mimeo). Faculty of Economics, University of Cambridge, Cambridge, UK.

Nagubadi, R.V. and Zhang, D. (2013) U.S. imports for Canadian softwood lumber in the context of trade dispute: a cointegration approach. *Forest Science* 59(45), 517–523.

Parajuli, R., Zhang, D. and Kosman, K. (2018) Province specific impacts of the 2006 United States–Canada softwood lumber agreement: a seemingly unrelated regression approach. *Forest Policy and Economics* 90, 1–6.

Pesaran, M.H. and Shin, Y. (1996) Cointegration and speed of convergence to equilibrium. *Journal of Econometrics* 71, 117–143.

Pesaran, M.H., Schuermann, T. and Weiner, S. (2004) Modelling regional interdependencies using a global error-correcting macroeconometric model. *Journal of Business and Economics Statistics* 22, 129–162.

Pesaran, M.H., Schuermann, T. and Smith, L.V. (2009) Forecasting economic and financial variables with global VARs. *International Journal of Forecasting* 25(4), 642–675.

Smith, L.V. and Galesi, A. (2014) GVAR Toolbox 2.0. Available at https://sites.google.com/site/gvarmodelling/gvar-toolbox (accessed July 20, 2020).

Statistics Canada (2019a) Table 980-0044 Domestic exports - wood and articles of wood; wood charcoal (table) CIMT (database). Canada, Ottawa.

Statistics Canada (2019b) Table 36-10-0104-01 Gross domestic product, expenditure based, Canada, quarterly (table) CANSIM (database). Canada, Ottawa.

Statistics Canada (2019c) Table 34-10-0135-01 Canada Mortgage and Housing Corporation, housing starts, under construction and completions, all areas, quarterly. Canada, Ottawa.

U.S. Bureau of Economic Analysis (2019) SQGDP2 GDP by state 1/ GDP by state: all industry total. Available at: https://apps.bea.gov/itable/iTable.cfm?ReqID=70& step=1#reqid=70&-step=1&isuri=1 (accessed December 6, 2019).

U.S. Census Bureau (2019a) Building permits survey. Available at: https://www.census.gov/construction/bps/statemonthly.html (accessed December 6, 2019).

U.S. Census Bureau (2019b) Data relationships between permits, starts, and completions. Available at: https://www.census.gov/construction/nrc/nrcdatarelationships.html (accessed December 6, 2019).

U.S. Census Bureau (2019c) Definitions of division, new residential construction, and structure. Available at: https://www.census.gov/construction/nrc/definitions/index.html (accessed December 6, 2019).

U.S. International Trade Data (2019) States' HST trade flows. Available at: https://dataweb.usitc.gov/trade/ (accessed December 7, 2019).

9 Discussion

G. Cornelis van Kooten

University of Victoria, Victoria, BC Canada

Abstract
The role and purpose of trade modeling in forestry is highlighted, as is the need for greater attention to data needs. Directions for future research are discussed.

Legislators are interested in knowing how forestry policies will affect the three areas that concern them the most – employment, revenues, and foreign exchange. The types of surplus areas defined in Chapter 1 are not generally considered of importance by policy makers, but not because they do not care about these things. Rather, welfare measures are not tangible, something that those affected can point to as a benefit or loss, particularly consumer surplus. Producer surplus (also referred to as quasi-rent) shows up as revenue over and above variable costs, deals with the survivability of firms, and, thereby, affects employment in particular and tax revenue over and above any revenue from stumpage fees. Yet, spatial price equilibrium models seek to maximize these surpluses; indeed, spatial price equilibrium (SPE) models seek to maximize the total surplus accruing to all regions and/or countries, not only that of one region or country. Employment, revenue, and even foreign exchange earnings are tied to harvest levels and production of wood products. Essentially, therefore, legislators would prefer to maximize the output of forest products subject to various environmental and other objectives of government.

It is important to recognize that anti-dumping and countervailing import duties, as well as policies such as log export restrictions, are the result of lobbying or rent seeking by private woodlot owners and producers of wood products. In the case of the Canada–U.S. softwood lumber dispute, this would be the U.S. Lumber Coalition. Unless U.S. Coalition members are able to capture the policy-induced economic rent that restrictions on imports would cause, they do not care if import tariffs, export taxes or quotas are chosen to reduce imports. Import tariffs would be collected by the importing government, export taxes by the foreign government, and quota rents by the producers

10.1079/9781789248234.0009

in the foreign country. The only benefit to U.S. producers of restrictions on foreign imports comes from higher prices for wood products. This would increase the quasi-rent (producer surplus) accruing to timber processing companies, although it would also lead to higher stumpage costs (as demonstrated in Chapter 3, this volume). Any restrictions on imports would cause forestland values to increase. That is, the benefit of restrictions on imports will ultimately get capitalized in land values.

Much the same can be said about log export restrictions. Log export restrictions reduce the price of logs to the benefit of domestic wood processors; however, the overall objective of log export restrictions is to increase employment in the processing sector. This comes at a cost: log export restrictions reduce domestic timber land values and lead to global inefficiencies in the production of wood products.

As discussed in Chapter 2, Canadian forestland tenure arrangements tend to result in provincial control of harvests. Provincial governments set harvest levels, not in response to market prices for wood products, but to prevent declines or even large fluctuations in employment in forest operations and at mills, and thereby ensure the (economic) stability of rural communities. This is despite evidence indicating that such policies do not mitigate instability. Nonetheless, the implication of this policy is that stumpage fees have little impact on the output of post-harvest wood products, although firms have some flexibility to shift harvests across years as long as they meet harvest targets over a 5-year period. The implication is that, even if Canadian lumber producers are advantaged by lower stumpage fees than would be the case if timber was sold by auction, it does not lead to higher levels of output and exports. The only impact of lower than market-determined stumpage prices is a reallocation of resource rents from government to wood processors, forest-sector workers (via above-average wages), and/or rural communities by (somewhat) reducing business cycle fluctuations. Resource rents are essentially dissipated through economic inefficiency that occurs because too little timber is harvested on some occasions and too much on others. Tenure arrangements are ignored for the most part in the development of the models discussed in this book, particularly in the spatial price equilibrium models.

What, then, is the role of forest products trade models? Clearly, by linking production and exports of wood products to employment, trade models can provide decision makers with information regarding the effects that various trade restrictions could potentially have on employment (although this assumes a fixed ratio between wood product output and employment). Trade models can also provide notions of the effect of various forestry policies on potential revenues and foreign earnings.

Forest-sector trade models are also important for analyzing the effects of policies by countries that seek to prevent log exports or seek to subsidize wood biomass for producing electricity. As discussed in Chapter 6, SPE trade models have been used to investigate a variety of forest-sector policies. For example, regulations and subsidies by European countries to increase the use of wood pellets in lieu of coal to produce power have been analyzed by Johnston and van Kooten (2016) using the REPA forest trade model. The authors found that an increase in the demand for wood products would reduce global prices of lumber because sawmill residues are a major source of fiber for pellet production. This in turn leads to greater storage of carbon in wood products, although the prices of other wood products, including pellets, increased because those products compete for residual fiber. Likewise, Johnston and Radeloff (2019) used the GFPM to investigate the potential to mitigate climate change by storing carbon in post-harvest wood products.

One purpose of the current contribution is to encourage readers to consider SPE trade models for examining various global or country-level policies that affect the forest sector, locally and globally. The Canada–U.S. softwood lumber dispute has been important in stimulating demand for forest products trade modeling for more than four decades. Despite such a long history and some drawbacks to these types of models, there remain many issues that can still be fruitfully analyzed within this framework. The models provided here can best be considered as a jumping-off point for future work. The models themselves might, in some sense, be timeless; the challenge is rather that of finding, maintaining, and updating the databases that underlie all such models. In addition, it is necessary to determine, either from the literature or through empirical estimation, elasticities of supply and demand for each of the regions in a trade model. These challenges can only be met when stable funding of these endeavours is available.

Finally, more research is required into the construction of gravity and global vector autoregressive models. These offer linkages between the policies in other sectors and the forest sector. Again, such modeling efforts require maintaining and updating quality datasets. In this regard, the two types of models – spatial price equilibrium trade models and econometric constructs – feed off each other. Clearly, there remain many exciting opportunities for forest economists to continue to contribute to policy analysis while conducting frontier-type research in trade modeling more broadly.

References

Johnston, C.M.T. and Radeloff, V.C. (2019) Global mitigation potential of carbon stored in harvested wood products. *Proceedings of the National Academy of Sciences of the United States of America* 116(29), 14526–14531.

Johnston, C.M.T. and van Kooten, G.C. (2016) Global trade impacts of increasing Europe's bioenergy demand. *Journal of Forest Economics* 23, 27–44.

Index